Technical Debt in Practice

Technical Debt in Practice

How to Find It and Fix It

Neil Ernst, Rick Kazman, and Julien Delange

The MIT Press
Cambridge, Massachusetts
London, England

The MIT Press would like to thank the anonymous peer reviewers who provided comments on drafts of this book. The generous work of academic experts is essential for establishing the authority and quality of our publications. We acknowledge with gratitude the contributions of these otherwise uncredited readers.

This book was set in Stone Serif by Westchester Publishing Services. Printed and bound in the United States of America.

Library of Congress Cataloging-in-Publication Data

Names: Ernst, Neil, author. | Kazman, Rick, author. | Delange, Julien, author.
Title: Technical debt in practice : how to find it and fix it / Neil Ernst, Rick Kazman, and Julien Delange.
Description: Cambridge, Massachusetts : The MIT Press, 2021. | Includes bibliographical references and index.
Identifiers: LCCN 2020041281 | ISBN 9780262542111 (paperback)
Subjects: LCSH: Computer software—Development—Quality control. | Software failures—Prevention. | Project management. | Maintainability (Engineering)
Classification: LCC QA76.76.Q35 E76 2021 | DDC 005.3028/7—dc23
LC record available at https://lccn.loc.gov/2020041281

10 9 8 7 6 5 4 3 2 1

—For Kieran, Elliott, and Kambria, who toured our old hometown while we labored on this. N.E.
—For Hong-Mei, who humored our rants and cheered us on. R.K.
—For Alejandra and Chewie. J.D.

Contents

Acknowledgments

This book would not have happened without the help and inspiration of a lot of different people. The three of us met while working at the Carnegie Mellon University Software Engineering Institute in Pittsburgh. Thanks to all our coworkers at the SEI, including Robert Nord, Peter Feiler, Ipek Ozkaya, James Ivers, Felix Bachmann, John Klein, Stephanie Bellomo, Phil Bianco, and Chuck Weinstock. Linda Northrop remains an inspiration.

Rick Kazman would especially like to thank his research collaborators, including Yuanfang Cai, Damian Tamburri, Humberto Cervantes, and the group at SoftServe, including Serge Haziyev and Andriy Shapochka.

Julien would like to thank Julien Danjou and Nicolas Devillard for their time.

Neil would like to thank his collaborators, as well as the anonymous reviewers of this manuscript, who although causing more late nights, doubtless improved the final version.

1 Introduction

> There is scarcely anything that drags a person down like debt.
>
> —P. T. Barnum

1.1 What Is Technical Debt?

If you have a credit card or a mortgage or a car loan, you already know a bit about debt. You are also probably familiar with the many metaphors that surround the ways we think about and talk about debt: crushing debt, drowning in debt, buried in debt. We talk of debt in relationships, or the social debt that you feel after your neighbor invited you to dinner and you have not reciprocated. Metaphors are pervasive in our lives. So it should come as no surprise that we also apply this metaphor to the technical world. The fact that you cracked the cover of this book suggests that this metaphor is already speaking to you. The metaphor of technical debt is something that every software developer has at least heard of. But metaphors only take you so far. Our purpose in writing this book is to move us all beyond just the metaphor into actionable insights, methods, and tools that allow us to deal with technical debt as software engineers.

Every organization that creates nontrivial software systems has technical debt, and the software development world is becoming increasingly conscious of this debt. Take Facebook, for example: the company originally used the PHP language to prototype and deliver their product quickly as a young company, but then as they grew they had to face the limitations of their early technical decisions. PHP simply was not able to scale and provide the kind of performance that they needed to support their ever-growing userbase, and so Facebook had to find solutions to pay this debt back; in their case, the solution was to create a PHP transpiler. Almost all companies face such issues, and this book provides many concrete examples of technical debt—from Boeing and Airbus to Twitter and many others. But we do not just provide horror stories (although there are plenty). We provide a concrete set of practical solutions—methods, tools, and techniques—for dealing with technical debt. An important message of this book is not that you should never incur debt; it is that you should not *inadvertently* assume a large debt, at least not without knowing when and how you will pay it back. And we provide you the techniques to do just that: to identify debt, to manage the debt that you have identified, and to avoid debt where possible.

The metaphor likening the creation of software to a process of going into debt was invented by Ward Cunningham in 1992. Cunningham was inspired by the seminal work of the linguist and philosopher George Lakoff on the power of metaphor. Lakoff argued that metaphors are central to the development of ideas and that humans are almost unable to talk and reason without using metaphors.

Cunningham, inspired by this insight, created the debt metaphor as a way of explaining his software development actions and decisions to his boss. (Actually, Cunningham did not call it "technical debt"; others coined that term later. Cunningham just called it "debt" in his early writings.) His boss came from the finance side, and Cunningham knew that his best chance of getting his boss to understand *what* he was doing and *why* was by using a metaphor to link with something that he felt his boss would be able to relate to: going into debt. Cunningham noted that, in the world of business and in our personal lives, we often borrow money to achieve a goal more quickly (like buying a car or a home). But then we have to live with the "penalty" of this decision: we pay interest on that loan.

Cunningham thought that going into debt could be a good thing for the eventual state of the software product. The world agrees with him. As a society we love our mortgages, car loans, and credit cards. In the world of software, we try to get a product to market as quickly as possible, even though this product may be flawed, even though we may not fully understand our market and hence the features that we should be providing, and even though, in our hurry to get the product out the door, we inevitably take shortcuts and make some ill-considered decisions.

Despite all these negatives, Cunningham still maintained that "going into debt" was a *good* thing. For Cunningham, it was an essential part of building software iteratively, naturally occurring as your theory of the software diverged from what you had actually written. Paying down the debt meant realigning the implementation with the theory. Just as in real life, these loans need to be repaid. Debt is simply a means to an end. In real life, if you do not repay what you borrow, more and more of your income goes to servicing your debt. Your credit score drops, collection agencies appear at your door, and your ability to borrow further is greatly decreased. In the world of software, if we do not repay the debt by refactoring—repairing and improving the hasty and ill-considered code that we first delivered—then our debt may, and often does, overwhelm us. We will then spend all of our effort fixing bugs and never get to add new features, and those new features are what our customers really value, after all. (If you are an experienced developer you are probably saying "been there, done that" at this point.)

In software engineering, the perfect is usually the enemy of the good. If we wait until we have a perfect understanding of our requirements and a beautiful, polished design, we will have likely missed our window of opportunity to capture market share. We may have alienated our users and lost the support of management. A better way is to develop something, present it to users, get feedback, and iterate, gradually and incrementally improving the product as we learn more about the problem domain. So debt is not, by itself, the problem. The problem is not identifying and acknowledging the debt, not measuring the debt, and not paying the debt back. We must, in fact, move beyond the metaphor. We must not only acknowledge our debt, but measure it and manage it. But typically we do neither; we simply ignore it. Why does this happen?

This happens because, in the early stages of a product, the interest payments are not too large. The code base is growing, but often it is still comprehensible by its creators. We are adding features without too much trouble. We are happy and our customers are happy. We are getting positive feedback, which reinforces our behavior and our motivation: get the next feature out the door.

Over time, of course, complexity increases. We might keep up for a while: some people can keep a few thousand lines of code in their head—they can digest and assimilate all of this information and operate on this knowledge very efficiently. Some exceptional programmers might be able to keep 20,000 or 50,000 or even 100,000 lines of code in their head. But ask yourself: What is the largest system allowed by law? That is a stupid question of course. No laws constrain artificial systems; they have no inherent bounds. That is why the Chromium project has nearly 20 million lines of code and the Linux kernel is almost as big. Clearly no human can digest 20 million lines of code. If you could read one line of code per second and did this nonstop, twenty-four hours a day, it would take you nearly eight months to read the entirety of Chromium.

It is a fact: system complexity, if left unmanaged, will always overwhelm us. Thus, a system that starts off as comprehensible becomes incomprehensible over time, despite our best efforts. We can fool ourselves in the early stages of development that everything is OK, and that we can ignore the growing complexity—the mounting debt—because we *can* keep it all in our heads, but in the end it overwhelms us. That is, it overwhelms us unless we pay down some of this debt, unless we consciously attempt to manage the complexity, unless we forego some of the instant gratification that worked so well for us in the early stages of the project (delivering features) to do the less visible but critically important task of re-envisioning, restructuring, and refactoring our technical artifacts (e.g., code, tests, documentation). Part of this book is a series of case studies that explain exactly this process of moving through early to late phases of development.

Slowly, as the field of technical debt matured, people started to realize that there were many forms of debt and many nuances regarding how, when, why, and if one should take on debt. Martin Fowler created a nice taxonomy of technical debt back in 2009, which we have adapted slightly, as shown below in figure 1.1.

In this figure, Fowler distinguishes two dimensions of debt. Dimension 1 (which we call "risk") explores this question: Is the debt reckless or prudent? If one simply codes without ever considering design, that would be reckless. That would be equivalent to constructing a building without ever having a civil engineer or architect design or analyze it. This is unthinkable (and illegal) in the construction industry. On the other hand, debt may be strategic and prudent. For example, if we are facing a hard deadline, if we need to prototype and produce *something*, so that we can get feedback and figure out what we really need to build, these are cases where taking on debt is prudent. Or, to turn the argument around, spending a large amount of effort on design in these instances would be a waste, since we have more important goals (such as the deadline) or little confidence that we would be building the right product (which is why we prototype). There is no point in building the wrong

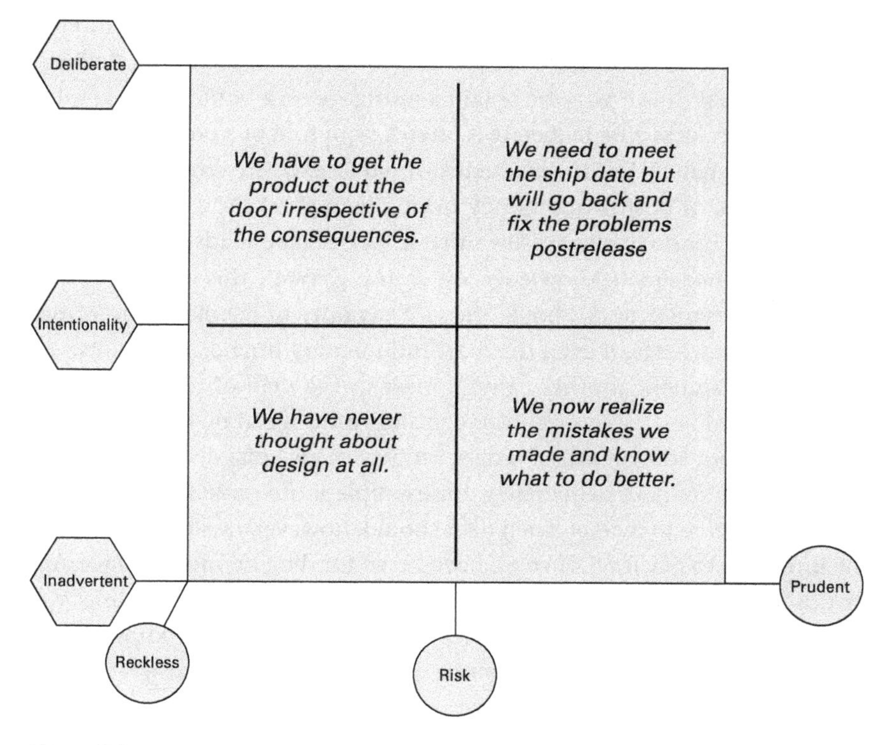

Figure 1.1
Technical debt quadrants.

product right, or in delivering a perfect product after the deadline has been missed. So taking on debt in such cases is perfectly reasonable if (and this is the hard part) we commit to analyzing and repaying the debt in the future.

Fowler's dimension 2—which we call "intentionality"—explores a second question: Is the debt deliberate or inadvertent? There are times when a debt is both deliberate and reckless, as scary as that sounds. This is often the result of management decisions: agreeing to unrealistic deadlines, understaffing, not hiring the right staff, or not training staff adequately. In such cases, a product is often thrown together while developers don't realize they are even taking on debt; perhaps this notion has never even occurred to them. They are simply in survival mode, trying to meet their deadlines and goals. However, there are also cases where the debt is deliberate. In this case, the team knows that they are taking shortcuts and they do so as a calculated investment. The argument goes like this: We need to make some suboptimal decisions now, to meet our near-term needs, but we fully intend to pay back the costs of these decisions. Debt that is prudent and deliberate is good debt; this is why we obtain a mortgage on a home, for example. Debt that is reckless and deliberate is often a symptom of a project that is in trouble. The analogy in financial terms would be excessive credit card debt, payday loans, or borrowing money from a loan shark. You really cannot afford it, but you do it anyway. Debt that is reckless and inadvertent is often the result of inadequate knowledge and training. While this likely does not apply to the readers of *this* book, the vast majority of people in the world have little knowledge of even the most rudimentary financial concepts.

Steve McConnell, another early pioneer in the field of technical debt, further refined and categorized the concepts introduced by Cunningham. He wrote about short-term vs. long-term debt. As we just discussed, sometimes we take on debt deliberately, for example to preserve startup capital or to reduce time to market. Such debt should, however, be short-term with the intention to pay it off once we have better funding or once we have our first product release in the marketplace. On the other hand, sometimes we take on long-term debt for strategic reasons. For example, if we know that a system is nearing the end of its lifetime, but we can still squeeze a few more years and a few more dollars out of it, then it likely makes sense to just make the minimal changes needed to keep it going. But note that in both cases, the decisions about taking on (and repaying) debt are intentional. Unintentional debt typically reflects poor knowledge, poor discipline, or bad programming practices, as we will discuss throughout the book. Sadly, most people are not adept at managing debt, and programmers are people (see the Financial Literacy Test sidebar in box 1.1 if you believe otherwise).

Box 1.1
Financial Literacy Test

About a decade ago Olivia Mitchell, a professor at the George Washington University, devised three simple questions to assess basic financial literacy. These were:

1. "Suppose you had $100 in a savings account and the interest rate was 2 percent per year. After five years, how much do you think you would have in the account if you left the money to grow?"

   a. More than $102
   b. Exactly $102

c. Less than $102
d. Don't know
e. Refuse to answer

2. "Imagine that the interest rate on your savings account was 1% per year and inflation was 2% per year. After one year, with the money in this account, would you be able to buy . . . "

a. More than today
b. Exactly the same as today
c. Less than today
d. Don't know
e. Refuse to answer

3. "Do you think the following statement is true or false?

Buying a single company stock usually provides a safer return than a stock mutual fund."

a. True
b. False
c. Don't know
d. Refuse to answer

Shockingly, just one third of Americans age fifty and older were able to correctly answer all three of these questions. And these results held up in further testing done in many other countries. While the answers improved slightly as the level of the respondents' education went up, even these results were still depressing. Only 44.3% of respondents with a college degree answered all three questions correctly.

What is the point of this story? It is quite simple. If we humans are so bad at thinking about financial debt, we are even more disadvantaged when thinking about technical debt. If even the simplest, most basic knowledge of financial literacy is uncommon, how likely is it that the average programmer, architect, project manager, or nontechnical manager would understand concepts of debt as they apply to the more abstract world of code and design, where the consequences of debt are often not felt for years? Thus, the point of this aside is to remind you that understanding debt and its consequences is likely not going to be intuitively obvious to most members of your organization. You are going to have to do some work to build a business case for monitoring and, in some cases, paying back, your accumulated technical debt.

Oh, and by the way, the answers to the financial literacy quiz are A, C, and B. If you got any of these wrong, you need to fix your financial literacy!

1.2 Moving Beyond the Metaphor

While technical debt began as a metaphor, it is slowly morphing into something that can be managed. If we want to believe that, as software engineers, we are truly engineers and not merely hackers, we need to measure, track, and reason about our software. In her landmark article "Prospects for an Engineering Discipline of Software," Mary Shaw outlined what it means to be an engineering discipline. She defined engineering as: "Creating cost-effective solutions to practical problems by applying scientific knowledge building things in the service of mankind." Engineering is based on a scientific foundation. To properly manage software engineering processes and artifacts, they must be measured. As any management scientist will tell you, if you can't measure it, you can't manage it. And yet the vast majority of software built today is not designed, not measured, and not managed. Because the demand for software is ever-growing and because software is so malleable, we have (mostly) been able to get away with this cavalier attitude. But make no mistake: this is not engineering. Software development, the way it is practiced in most projects is, at best, a craft. And a craft depends on the skill of the engineer.

As we will show in this book, there is hope: we can move beyond the metaphor of technical debt. We can move beyond software engineering as a craft. But doing so requires a change of habits, it requires a change of tooling, and it requires discipline.

The technical debt metaphor is growing to encompass more aspects of a modern software project, beyond source code. For example, there are now tools that can measure and analyze social debt and design debt. Let's consider social debt. There are numerous examples of projects that have been driven into the ground not because their technology was inferior but because their organizational structure was pathological. Just as researchers have identified code smells—duplicate code, god classes, inappropriate intimacy, and so on—a number of organizational smells have been identified. For example, Tamburri and colleagues have identified organizational smells—social debt—such as:

- *Cognitive distance*: the distance developers perceive among peers with considerable (educational, experiential, cultural) background differences.
- *Informality excess*: excessive informality due to the lack of information management and control protocols.

- *Disengagement*: thinking the product is mature and deploying it even though it might not be ready.
- *Institutional isomorphism*: excessive similarity of the processes and structures of one subgroup to those of others.

Companies are beginning to invest in monitoring and managing technical debt. For example, in one company we analyzed eight projects and applied three complementary analysis techniques to measure and monitor their technical debt. In chapter 4, we describe an analysis of the code bases, issue-tracking systems, and version control systems of these eight projects to assess the quality of their software architectures. The outputs of the analyses were (1) architecture-wide coupling measures (decoupling level [DL], and propagation cost [PC]) that the company could use to compare their systems to industry benchmarks, (2) a set of design flaws found in each project, and (3) a set of design "hotspots" in each project—groups of files that architects and developers consider to be worth refactoring.

By measuring and tracking these eight projects, and by quantifying the costs incurred by the accumulated debt, we were able to convince management to refactor six of them. Prior to our collecting this data, the project members were vaguely aware of their debt—they of course experienced it every day—but they had no way to measure it and hence no way to convince their managers to invest the time and effort to pay down the debt. A connected metaphor comes (initially) from the world of automotive engineering. *Lean thinking* suggests that one important practice in manufacturing and producing software is to reduce the amount of waste. Waste is any practice that is not directly adding value to the product. Technical debt is waste, either in the sense of extra work to deal with the effects of the debt or as required rework to remove the debt that we should never have incurred in the first place (and, ultimately, have financial and productivity impacts).

Thus, this book is not a depressing tale but rather a reason for hope. The good news is that we should, and we *can*, explicitly monitor and manage technical debt.

1.3 Summary

Our message is that software development and technical debt can be controlled and managed. But this process of managing must begin with measurement.

This is the message of the remainder of this book: technical debt is real, and it has enormous consequences on project success. For this reason, a prudent project manager or technical lead needs to plan for how to avoid, detect, monitor, manage, and pay down debt.

In the rest of this book, we will show how you can have debt in your design, in your code, in your test suites, and in your documentation. And we will provide practical strategies in each case for avoiding this debt. But that is the ideal case. In many projects debt is already there, or it is unavoidable due to schedule pressures. To help resolve existing debts, we also provide strategies for identifying and quantifying this debt and for making the business case to pay down the debt.

1.4 Book Outline

The majority of the book is dedicated to the details of dealing with technical debt in the software development lifecycle. We include chapters on requirements debt, implementation debt, testing debt, architecture debt, documentation debt, deployment debt, and social debt—broadening the horizons of what has been typically thought of as technical debt. In each of these chapters we make the case for why each of these kinds of debt is worrisome. Figure 1.2 briefly highlights how these different lifecycle activities can cause technical debt. This is a big-picture view and as such is incomplete, but it is suggestive. We examine every major project activity—requirements, design and architecture, testing, deployment and delivery, and documentation—and their consequences for technical debt. We also introduce the little-studied but pernicious topic of social debt.

This characterization of technical debt may be different from what you have read in other books or articles. Most of the existing literature and most technical debt detection tools focus on *code* as the carrier of debt. While we agree wholeheartedly that code may contain considerable debt, you can "go into debt" in other dimensions of your software project just as easily. After all, debt is incurred whenever you took a shortcut, or whenever you did not understand or manage an aspect of your project lifecycle as well as you could have. For example, if you have a large team and subparts of the team are not communicating with each other, this might be a problem for your project. It is a debt because you might have avoided this by paying more attention to mentoring new project members or fostering better

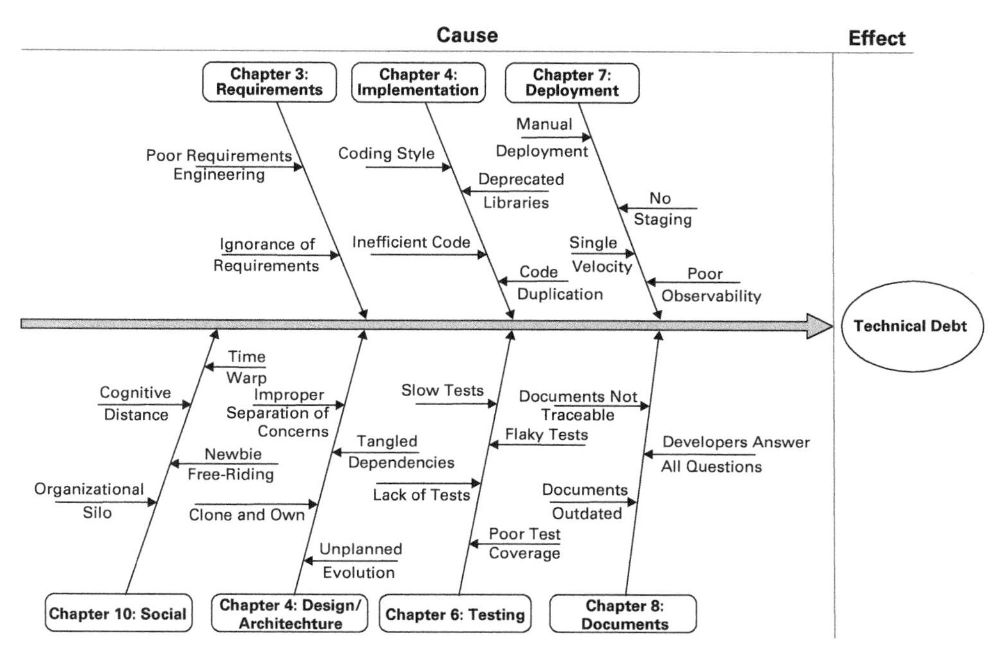

Figure 1.2
Fishbone diagram of the causes of technical debt, per software lifecycle phase. A fishbone (or Ishikawa) diagram represents a problem on the right (head) and causes of that problem on the left.

communication channels and practices. But if you, instead, just let everyone loose on the project and hoped for the best, you have incurred what we call social debt.

Let's consider another example: Perhaps your project started off as a simple codebase with a small team of developers who all knew each other well and worked closely together. As the project grew, however, a great many new developers were added, some of whom were working in remote locations. The old way of doing things did not require anyone to document anything; you could always just walk over to your buddy or chat over lunch and share information informally. However, as the project progressed and grew this was no longer feasible, and many people were spending a lot of time reinventing the wheel. In this case, you may have incurred documentation debt. If you had just spent some time documenting the rationale behind the system, coding standards, architectural assumptions, and design constraints you might have avoided this debt. But in the early days of the project this simply was not needed and was not a priority.

The many types of technical debt and their associated causes are not independent of one another, of course. Requirements debt leads to a poor understanding of what the software should be doing. That makes it hard to know what the architecture ought to be. And it makes it unlikely that anyone would bother to document these shifting sands. If the architecture is poorly understood, then the implementation will be prone to problems—and we will certainly have trouble writing tests if we don't know what the original requirements were. Finally, if we are incurring debt in all these phases, it is highly likely we will also have social debt that either causes or is caused by all of the other problems.

Thus we see the potential for debt in every aspect of the software development lifecycle, and we have organized our book accordingly, with chapters on requirements debt, implementation debt, architecture debt, testing debt, documentation debt, and deployment debt. Within each of these chapters we have adopted a common structure. First, we discuss how, in each lifecycle phase, technical debt can be **identified:** how the causes in figure 1.2 can be discovered, using techniques and tools and metrics. We then explain how to **manage** these problems, once they have been identified, and how they can be mitigated. Finally, to **avoid** the problems in the future, we discuss strategies to improve the conduct of that phase of the lifecycle so that technical debt will not reoccur. At the end of each chapter, we conclude with a section on further reading, where the sources for our discussion are elaborated.

We also examine the notion of technical debt in machine learning systems (chapter 9). As these systems are relatively new—compared with traditional forms of software development—we felt that it was important to evaluate the kinds of debt found there and the ways in which these new forms of debt can be identified and managed.

In the chapters on the social and managerial aspects of technical debt (chapters 10 and 11), we delve into more details about how projects, and project managers and architects, can deal with the inevitability of technical and social debt. We talk about how to measure code-level and programmer-level activities as indicators of such debt types. Social debt is the notion that many technical debt issues are caused by organizational problems. The technical debt metaphor lends itself to financial reasoning, and we discuss a few project management techniques for measuring and managing debt.

Throughout the book we intersperse our discussion of the various kinds of debt with a set of interviews with practitioners—detailing their experiences with various kinds of debts—and case studies from real projects, which were (or are still) affected by technical debt. We involve these practitioners in boxes we call "Voice of the Practitioner," where we distill lessons on managing technical debt from a practical perspective. We present the full interviews at the end of the book in an appendix. From each interview we summarize the purpose of the interview and some key technical debt takeaways. We do the same thing with our case studies, explaining how the case study explains and reveals key issues with technical debt.

In every chapter we provide concrete advice on how technical debt can be identified, avoided, and managed to deliver a higher-quality product faster than before. We give real-world examples of how teams have benefitted from debt removal and how they have been able to address long-term issues, such as performance problems, functional regressions, release delays, or design decay.

Further Reading

For a fascinating discussion of the use and power of metaphors, see: George Lakoff, Mark Johnson, *Metaphors We Live By* (University of Chicago Press, 1980).

The term "technical debt" (or, originally, just "debt") was coined by Ward Cunningham, in "The WyCash Portfolio Management System," http://c2.com/doc/oopsla92.html, 1992. This was later refined and elaborated by many others, such as Martin Fowler, where he described his "TechnicalDebtQuadrant," https://martinfowler.com/bliki/TechnicalDebtQuadrant.html, 2009; and Steve McConnell, in his essay "Technical Debt," http://www.construx.com/10x_Software_Development/Technical_Debt/, 2007. George Fairbanks summarizes the iterative nature of debt in his IEEE Software article "Ur-Technical Debt," https://www.georgefairbanks.com/ieee-software-v32-n4-july-2020-ur-technical-debt.

Annamaria Lusardi and Olivia Mitchell have discussed their extensive research and studies on financial literacy in "The Economic Importance of Financial Literacy: Theory and Evidence," *Journal of Economic Literature* 52, no. 1 (March 2014): 5–44.

Mary Shaw has written and spoken extensively on the progress of software engineering as a true engineering discipline. This was first presented in the technical report: "Prospects for an Engineering Discipline of Software," CMU/SEI-90-TR-20 (September 1990). The quote is found on page 2.

A discussion of social debt and how an architect may detect it and defend against it can be found in Damian Tamburri, Rick Kazman, Hamed Fahimi, "The Architect's Role in Community Shepherding," *IEEE Software* 33, no. 6 (November–December 2016): 70–79.

Lean thinking was introduced to North American audiences in Eliyahu Goldratt and Jeff Cox, *The Goal: A Process of Ongoing Improvement* (North River Press, 2014). Donald Reinertsen has pioneered product development and the concept of cost of delay in "Principles of Product Development Flow" (Celeritas, 2009). The notion of applying Lean to software was introduced in Mary Poppendieck and Tom Poppendieck, *Lean Software Development: An Agile Toolkit* (Addison-Wesley Professional, 2003). Expanding beyond software development, Eric Ries's book *Lean Startup* (Crown Publishing, 2001) has been influential in understanding how to experiment and pivot an entire company.

Finally, Cai and colleagues have substantial experience in automatically analyzing architectures for design problems that lead to technical debt. The foundation for this work can be found in Yuanfang Cai, Lu Xiao, Rick Kazman, and Ran Mo, Qiong Feng, "Design Rule Spaces: A New Model for Representing and Analyzing Software Architecture," *IEEE Transactions on Software Engineering*, January 2018. They outline a set of tools for automatically detecting design degradation and design flaws, and they provide empirical results showing the consequences of those flaws in Ran Mo, Yuanfang Cai, Rick Kazman, Lu Xiao, and Qiong Feng, "Decoupling Level: A New Metric for Architectural Maintenance Complexity," *Proceedings of the International Conference on Software Engineering (ICSE) 2016*, Austin, TX, May 2016; and in Ran Mo, Yuanfang Cai, Rick Kazman, Lu Xiao, "Hotspot Patterns: The Formal Definition and Automatic Detection of Architecture Smells," *Proceedings of the 12th Working IEEE/IFIP Conference on Software Architecture (WICSA 2015)*, Montreal, Canada, May 2015.

2 The Importance of Technical Debt

> Rather go to bed without dinner than to rise in debt.
> —Benjamin Franklin

We can think of the importance of technical debt in two ways. First, we should consider if we are incurring technical debt at a problematic rate. A problematic rate can be too fast, that is: "Are we adding to our debt burden?" But it could also be too slow: "Are we taking prudent risks?" Second, we need to decide if our code base already has high levels of technical debt: "Are we already drowning in debt? Is it preventing us from moving as fast as we want?"

2.1 Incurring Technical Debt

An emerging view of successful software development, popularized in lean methodology books such as Eliyahu Goldratt's *The Goal,* Eric Ries's *Lean Startup,* or Mik Kersten's *Project to Product,* is that a high-functioning software process *must* make mistakes. For many software projects the complexity of the problem is such that it would be impossible to understand the best process without experimentation. Experimentation has to involve releasing the software to production environments. A production environment is one where customers get to use the product. This means that some technical debt should be incurred, because if you aren't taking on debt you are either moving too slow or not learning all the lessons you need. Incurring debt in this way—that is, making a decision that results in technical debt—we refer to as *principal.* The principal is often invisible to end-users but all too visible to the system's developers. If you remember how Cunningham thought of debt, it was as a way to (a) get a working product into the real world as soon as possible but (b) had to be repaid as the gap between what was intended for the product and what it actually became grew larger.

When this gap gets too large, the effects of technical debt principal become visible to end users, in the form of bugs, quality problems, and software process slowdowns (bugs take longer to fix, new features longer to release). This visible form of debt is called *interest.*

The fact that technical debt might be incurred does not mean you should incur inadvertent/reckless debt (figure 1.1). Being careless is not the same as doing deliberate experimentation. Moving fast does not mean simply pumping out lines of code but rather delivering working, valuable software. There's even a school of thought, stated by Michael Feathers (see Further Reading), that argues that more code is actually a bad thing since it must all be maintained, and bugs are strongly correlated with code size. We shouldn't reward volume but instead units of value.

Sometimes deliberate and prudent debt is the best business strategy—deliberately recognizing that it is better to make *some* attempt at a solution than to strive for a *perfect* understanding that may never arrive. We do not want to over-control what is often the messy and imperfect process of building software. Although it is in the context of profiling and performance optimization, Donald Knuth's quote is apposite:

> The real problem is that programmers have spent far too much time worrying about efficiency in the wrong places and at the wrong times; premature optimization is the root of all evil (or at least most of it) in programming.

Premature optimization is no less of a problem in the context of software design than it is in programming. A smart strategy, therefore, is to focus on the important aspects of the problem as you understand them today, and release some version of the system to get feedback and to get a sense for what it should be doing. Once you update your mental model of the problem (perhaps you should have used a linked list instead of an array, for example, or you need to optimize retrieval of data but not storing data), version two can reflect this deeper understanding. But indeed the delta between versions one and two was the removal of the technical debt you borrowed to get version one released. The reason the debt made sense was because it wasn't until version one was finished that you could understand the better approach.

A great example of using technical debt deliberately is from the initial technical architecture of Facebook. Facebook was first written as a small PHP website (FaceMash) for rating the appearance of female students at

Harvard. Doubtless Mark Zuckerberg realized PHP wasn't the most elegant of web languages; equally certain is that choosing a more elegant, esteemed language may have dramatically impacted FaceMash's ability to get a product to market quickly and to grow the product. Here, usability and features were the priority. As a result, the original Facebook product code was debt-laden. PHP and the Zend web framework could not support Facebook scale as the user base grew. Facebook paid down technical debt by writing a transpiler, HipHop, to take the PHP code and turn it into C++. This reduced CPU load by half and served requests from the browser two and a half times faster. However, the resulting C++ codebase was five times the size of the original, implying there remained plenty of debt.

Paying down debt does not mean no debt remains. It should mean that the constraints (interest) the original debt imposed have been removed. Facebook's early growth is an excellent example of incurring technical debt, and it showcases the value and cost associated. The key insight is to ensure you or your company understand what is being incurred and whether it is worth taking on.

2.2 Living with Technical Debt

Does it matter if our code base contains technical debt? Of course! But only if we are feeling the impact, by paying down *interest,* or anticipate feeling the impact. Consider the happy situation of a bank holding your mortgage, then going bankrupt (assuming they cannot assign your mortgage). This debt need never be repaid, and thus was not worth worrying about.

Similarly, many codebases contain many areas with little to no attention being paid to them. Perhaps that part of the system is supporting a lesser feature that has little priority at the company. Perhaps it is a third-party supported feature that is in the repository but not actively developed. Technical debt in any part of the codebase that is causing real pain (interest payments in the form of reduced productivity, increased bugs, and so forth) should be prioritized for removal, just like any feature would be prioritized for implementation. See chapter 11 for tips on prioritizing technical debt.

Given finite resources, the debt we should focus on is located in more frequently changing, complex parts of the code. If we rarely need to change a part of the code, fixing it might be wasteful and an opportunity cost, since our developers could be doing other things. We care about technical debt because when found, it indicates parts of our codebase where large, lurking problems live, and the technical debt principal needs repaying.

How do we identify it? That will be covered in detail in later chapters, but one symptom is when we have indications of technical debt interest payments: historical data pointing to frequent bugs, large amounts of churn, and high design complexity. Technical debt indicates serious implications for the long-term quality of our code, and more importantly, implications for our ability to maintain a competitive release pace.

2.3 Practical Consequences of Technical Debt

We now present two brief anecdotes about the importance of paying attention to technical debt. The first story is about an initial release of a product where technical debt was so high that rework was required even before its launch, and consequently deadlines were missed. The second example is of ignoring technical debt too long, until a big-bang rewrite was required.

Story 1: The Phoenix Pay System

The Phoenix pay system was a payroll project created for the Canadian federal government based on a heavily customized version of PeopleSoft, implemented by IBM. The government pay system was quite complex, with hundreds of thousands of employees, numerous life events, complicated pay schemes (overtime, shift work, danger pay, etc.). Despite this complexity, the government insisted on scrapping the old systems, and it rolled out the Phoenix system in a big-bang fashion. Immediately problems were found:

- People on long-term leave found they were not paid their leave salary.
- People working overtime found the amounts to be incorrect in nearly all cases.
- People were overpaid, forcing them to store the money in a separate account for when the money was reclaimed. Subsequently the Phoenix system issued an income tax receipt showing this large amount, and income tax was charged as well (requiring an amended tax return when the system was eventually corrected).
- Public sector unions in response aggressively demanded immediate fixes and compensation, at taxpayer expense.
- The staff savings promised disappeared as new teams were created to deal with the mess.

Although not bug-free, the design and implementation debt in the Phoenix system was not the main issue, since it was a customized version of an existing product. Instead, as reported in a postmortem given by the Auditor General of Canada, "it was clear that the government had spent many years developing extremely detailed requirements that specified exactly what a private sector company should do if it won the contract. [The contractor] did precisely what the government specified, responding to multiple requests for changes throughout the length of the contract." However, the specification itself revealed little understanding of the fundamental issues (requirements debt), and the software was poorly tested (testing debt). Deployment and the management of the project were also characterized by short-term expediency.

Some have insisted that technical debt in software should only apply to design and implementation decisions that are not visible to end users. Our view, characterized by the Phoenix payroll example, is that the concept of short-term expediency that jeopardizes long-term success applies to a much wider set of software project problems. While technical debt in the software itself is important, the issues surrounding the software are equally important: releasing in an agile fashion, incrementally; small, frequent deployments that reveal requirements problems; clear understanding of business goals and objectives. In chapters 10 (social debt) and 11 (business concerns), we discuss these characteristics in more detail.

Story 2: Netscape 6

Netscape 6 was a major, near-total rewrite of the Netscape product, which was an early web browser. (As an aside, many of these war stories occur when a major rewrite is in the cards for a product.) Netscape had spent many years as the preeminent browser on people's computers, but by 1998 Microsoft and Bill Gates had famously realized the importance of the Internet. Bundling Internet Explorer (IE) as part of Windows meant IE's market share was increasing rapidly. Netscape Navigator, however, was a chaotic mixture of legacy code from the early era of the web: "We've pulled more useful miles out of those vehicles than anyone rightly expected," said lead developer (and Javascript creator) Brendan Eich. Consequently, the Netscape leadership decided to completely scrap the existing codebase. A new rewrite was started, called Mozilla (now known as Firefox).

The primary focus was to force the entire product to use a highly componentized architecture. For example, the web rendering component, called

Gecko, was common to the browser and the email client. However, the rewrite quickly bogged down in the complexity of a more modern browser client and the ambitious architecture approach chosen. For example, a component called XUL (XML User Interface Language) was intended to simplify cross-platform user interfaces, but it ran into challenges mapping to native code. Mozilla had also moved to a new (for the time) open-source model and, combined, those challenges made the new project slow to release its first version, further eroding customer confidence and market share.

One of the lessons of the Mozilla project was that continuing to coast on existing code, living with technical debt without fixing it or upgrading it, is a major risk. The second lesson is that technical debt typically cannot be resolved in a big-bang approach, at least not for complex software systems. In addition to identifying the debt your codebase has, some systematic approach for reducing the debt is important—a payment plan if you will. Facebook's upgrade to a PHP transpiler, which we mentioned above, happened in phases, and the approach allowed for front-line developers to continue to write the PHP code they were familiar with.

2.4 Managing Technical Debt Is Important

As we will see in the remainder of the book, the best approach to managing technical debt is to manage it proactively. This means making deliberate and justifiable decisions to incur debt, having a reasonable strategy for measuring and monitoring it, and paying it off in the future. Unsurprisingly, companies that build software as their main product tend to manage technical debt the best, using the most disciplined approaches. At Google, teams are routinely given time to refactor code to pay down debt and improve the quality attributes of the system—performance, modularity, security, and so forth. In many tech companies, technical debt is taken seriously and teams allocate time periodically to address outstanding issues and make sure that the code stays clean, up to company standards and the level of debt is kept under a given threshold. The very best companies understand that clean, high-quality, and maintainable code is a vitally important asset.

By contrast, we have found the most significant technical debt problems tend to occur in companies and organizations where software is not seen as a core corporate asset. For example, government contractors who until recently made most of their income selling hardware had little knowledge of software issues. The software now, however, makes up a majority of total system cost. A common approach in the contracted software world is "prototype as product." This happens when the bid team develops a killer prototype, wins the bid, and then hands the debt-laden prototype off to the development team to fix (which rarely happens). Debt is managed there in the Netscape model—wait until it is too late to fix it.

2.5 Future Proofing Technical Debt Is Hard

Technical debt is more than just badly written code. It refers ultimately to a mismatch between what the software should have been and what it actually is. One can have perfectly good code that worked well in one context (and time period) but is now totally unsuited to its intended purpose. An early IBM employee, Manny Lehman, noticed this as part of the IBM OS/360 project (as did Fred Brooks). He coined a set of laws that, taken together, boil down to the need to constantly maintain and update the software. That is not to say that all software should be designed for a multidecade service life (although see Case Study C for an example of how this might be done). Nonetheless, the management plans of software projects must be aware that the software will have to change and balance between over-engineering now (you aren't going to need it, or YAGNI) and a future system incapable of adaptation.

Poor coding style is certainly an indicator of technical debt, but we have found that the majority of truly problematic code is not detected by tools such as linters or static analyzers. These problems are often subtle, and it requires some detailed examination, data collection, and analysis before you can confidently conclude that there is technical debt. For example, the decision to use a particular version of a Raspberry Pi board for a prototype project might be eminently reasonable at the time. However, someone may record that this particular board has a modem that only supports 802.11b and not anything newer. This fact is duly forgotten, and a year goes by. The IT team upgrades the routers, and suddenly the product does not connect to Wi-Fi anymore, resulting in a complex, costly, and high-stress fix.

2.6 The Benefits of Technical Debt

Technical debt is useful leverage. In software development, technical debt, when deliberately incurred, is typically done to buy development speed—time to market. This is because the *value* of a particular release (the y-axis in figure 2.1) is composed of value of functionality and quality attribute properties of the code. But the net value also includes costs such as: the cost of a delayed release, the cost of implementation and deployment, and finally, of most interest to us, the cost of rework. Meeting a Christmas ship date for a game company is incredibly important, to the extent that games often ship with many known (even critical) bugs. Thus incurring the debt, in these instances, is demonstrably worth the boost in development speed that this choice brings. This is also known as balancing the cost of delay against the cost of rework (where this rework is typically fixing the bugs and smells that you included when you shipped). Such tradeoffs are common in the gaming world but also where quarterly deadlines or key consumer dates matter (e.g., fiscal year-end).

Figure 2.1 shows a notional software value curve. The y-axis indicates potential value as a function of functional value, nonfunctional quality attribute value, the cost of delay, the cost of rework, and the cost of each increment. The x-axis is time. As we continue to work on the codebase, value should be increasing over time. However, incurring debt eventually leads to interest payments such as extra bugs and the extra effort to find and fix bugs and to make changes. This all requires rework and refactoring, increasing costs, and reducing the value of the software. The difference between the hypothetical perfect trajectory (from point 2 to 4 and beyond) and the debt-burdened trajectory (from point 2 to 3 and beyond) is the amount of debt in the software.

2.7 Debts You Don't Have to Repay Are Good to Incur

Finally, technical debt only matters if it must be repaid. It is possible that the cost of the debt is close to zero if the bugs or additional complexity do not have any real, tangible impact. For example, part of the codebase may rely on an old piece of code to retrieve data from a mainframe. Perhaps in the ideal world this code would be upgraded, but if it works well, and has worked well for several decades, "fixing" it makes little sense. Of course, making this kind of determination ahead of time is not easy. Perhaps the mainframe is scheduled to be decommissioned in a year, and the software will no longer work. Understanding the tradeoffs behind this decision is the focus of (most of) the remainder of this book.

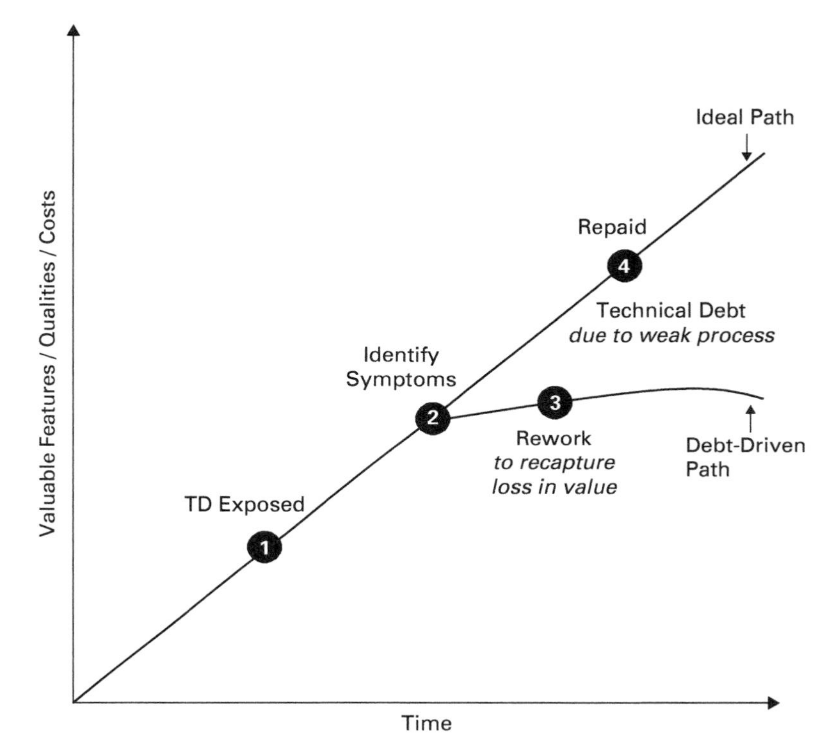

Figure 2.1
Technical debt and the software value curve.

2.8 Summary

In this chapter we discussed the importance of technical debt. Technical debt is part of your software process. It is incurred in the process. Poor process can lead to more debt. Similarly, technical debt is also part of your product. It can be detected and its removal prioritized like any other work (e.g., bugs, features, vulnerabilities). We learned that technical debt has been a factor in many system failures or problems, so you are not alone. Technical debt is incurred when we do not think about the future implications of our decisions, be they poor coding, bad design, requirements, testing, or lack of documentation.

Further Reading

We talked about Ward Cunningham in chapter 1: see Further Reading in that chapter for more details.

Knuth's quote comes from his Turing Award lecture, which is worth reading in its entirety: Donald Knuth, "Computer Programming as an Art," *Communications of the ACM 17*, no. 12), (December 1974): 667–673. The quote can be found on page 671.

The early history of Facebook is described by Katharine Kaplan, "Facemash Creator Survives Ad Board," *Harvard Crimson*, November 19, 2003, https://www.thecrimson.com/article/2003/11/19/facemash-creator-survives-ad-board-the/.

The Facebook PHP migration is described on the Facebook engineering blog (https://www.facebook.com/notes/facebook-engineering/hiphop-for-php-move-fast/280583813919/) and this paper at OOPSLA (https://research.fb.com/publications/the-hiphop-compiler-for-php/). The Mozilla big bang rewrite is first introduced in this roadmap: https://www-archive.mozilla.org/roadmap/roadmap-26-Oct-1998.html, and a retrospective a year later from Brendan Eich in 1998 can be found at: https://www-archive.mozilla.org/roadmap/roadmap-26-Oct-1998.html (search for "useful miles"). An example of how computer games defer hard technical problems to postlaunch is Battlefield 4's buggy launch: https://www.usgamer.net/articles/8-games-with-completely-broken-releases.

Joel Spolsky wrote eloquently about why refactoring (to pay down debt) is almost always preferable to rewriting: https://www.joelonsoftware.com/2000/04/06/things-you-should-never-do-part-i/.

One of the best conceptualizations of what programming and software engineering are is found in Peter Naur's idea of theory building, which he wrote about in Peter Naur, "Programming as Theory Building," *Microprocessing and Microprogramming* 15 (1985): 253–261.

Manny Lehman and Laszlo Belady, working on the same IBM 360 project as Fred Brooks, derived laws of software evolution, including the "law of continuing change" predating the agile movement by decades. See Meir Lehman, "Programs, Life Cycles, and Laws of Software Evolution," *Proceedings of the IEEE* 68, no. 9 (1980): 1060–1076, for some of the other laws.

The postmortem of the Phoenix pay scandal was covered in many Canadian news sources, but the official account can be found in a report by

Canada's equivalent to the Government Accountability Office (GAO), the Auditor General, published in 2017: http://www.oag-bvg.gc.ca/internet/English/parl_oag_201711_01_e_42666.html. The Treasury Board report, produced by GGI, is also useful: https://www.canada.ca/en/treasury-board-secretariat/corporate/reports/lessons-learned-transformation-pay-administration-initiative.html.

For more on code as inventory, Michael Feathers has captured the view in a nice, short article here: http://michaelfeathers.typepad.com/michael_feathers_blog/2011/05/the-carrying-cost-of-code-taking-lean-seriously.html.

3 Requirements Debt

> The two most important requirements for major success are: first, being in the right place at the right time, and second, doing something about it.
>
> —Ray Kroc

> Walking on water and developing software from a specification are easy if both are frozen.
>
> —Edward V. Berard

A requirement is a statement from a system stakeholder about some desired features or properties that the system should provide. Requirements come in various shapes and sizes. They can be issues in a backlog or statements in a lightweight text document. They can be modelled in a tool specifically for that purpose and linked with thousands of other artifacts in the organization. In general, we hope the requirements will not just exist in someone's head—but that often occurs as well.

Requirements are the link between the business or organizational value your company has and the technical, software implementations to realize those goals. Value for customers and end users is identified in a requirements *elicitation* phase (for example, by a marketing team, perhaps employing focus groups). The units of value identified in this phase are the requirements for the next release of the software. Increasingly, these units of value, these requirements, are traced all the way through product design and implementation, and eventually into production systems. In this way, a feedback loop exists from the results of A/B testing in the wild, all the way back to the initial focus group or ideation phase. This feedback allows the organization to track what works and what doesn't.

Each requirement should capture, succinctly, the initial idea for the new system feature or property that the business cares about. They are a means to support design discussions, track progress to a milestone, and discuss with customers the current state of their requests. We should be clear that requirements don't necessarily mean *requirements specifications*, although in some settings these are also necessary.

So what is technical debt in requirements—that is, requirements debt? One way to understand the relationship between technical debt in requirements and technical debt in design or code is by looking at the relationship between product value and intrinsic software quality. If the product is not delivering value to the client, then high intrinsic quality (i.e., low technical debt) is irrelevant. If intrinsic quality is low (i.e., the product has a lot of debt) then the likelihood of easily delivering valuable products in the future is likewise low.

3.1 Identifying Requirements Debt

3.1.1 Sources of Requirements Debt

Technical debt in requirements is incurred when we prioritize requirements for implementation that do not add value to the product. Most of the time this is due to these requirements becoming features that customers and users ultimately never use or care about (and therefore, are not willing to pay money for). For example, evidence suggests only 20% of the features in a given piece of software account for 80% of the use. That implies the remaining 80% of the features were largely unnecessary. Prioritizing requirements means selecting them for implementation to build a particular instance of a product. In an agile model, this might mean putting these requirements on the backlog. In acquisition settings, this might mean adding these requirements as must-haves in the formal call for proposals phase of the project.

Prioritizing one set of requirements also implies that other requirements are ignored or not implemented. So part of technical debt in requirements derives from ignoring important requirements (like regulatory requirements) that end up being important.

Often requirements debt happens because of inadequate or poorly conducted requirements elicitation and analysis, or because someone simply accepted any and all customer requirements without discussion. Debt incurred in the requirements phase refers to tradeoffs on what requirements

Box 3.1
Voice of the Practitioner: Andriy Shapochka

> The real debt causing most friction in the project is not captured in the tools or documentation in many cases. I think the most useful quantification of the technical debt could be represented as a relative cost of change required to evolve a specific part of the system to make it support new requirements, changing requirements, postponed requirements (can be functional or non-functional). The word "relative" here means comparison of what it takes to change the actual code with the would-be cost of change [under the] assumption the code is free from the identified technical debt item(s).
>
> —AS[1]

the development *ought to* prioritize. If requirements later change, we must update the design; but it is possible that this update would not have been necessary had we correctly predicted what requirements would be needed in the future (see box 3.1). For example, adding a new streaming video provider to a television could be done wirelessly, if this auto-update was an original function. Otherwise, the television either cannot handle this new service, or it needs to be updated manually, like with a USB card.

What is the role of technical debt in the requirements process? There are two main types of requirements debt:

1. *Poor requirements engineering:* this happens when we take shortcuts in requirements elicitation. For example, if we fail to talk to all the pertinent stakeholders, the resulting requirements will be of low quality or their prioritization will be incorrect. The result of this requirements engineering is thus often not useful for the system's stakeholders. This is a failure of the process of requirements gathering.
2. *Ignorance of requirements*: implementing features badly due to a poor understanding of the actual customer needs that are motivating the work. This is a failure of the process of software development.

Poor requirements engineering, in addition to causing pain in the requirements process itself, can cause or magnify the technical debt downstream. We should distinguish between the known, the unknown, and the unknowable. Some requirements are *known*: building a web application with no understanding of privacy requirements (such as the EU's General Data Protection Regulation, GDPR) is not radical experimentation but carelessness: these privacy requirements are known and now belong to the web-application design knowledge. Some requirements are *unknown*: we may not know that there is an applicable financial regulation (for example, related to some countries or states in which you operate), but we can discover this. However, some, perhaps most, valuable software requirements are *unknowable*. This means they have value that is only visible when customers and users actually get to play with them. In other words, they are unknowable ahead of time. For this reason, we need to ensure that features *not* implemented (i.e., unsatisfied requirements) are not debt. Later on in this chapter we explain how to distinguish iterative development from shortcuts in requirements and business analysis.

3.1.2 Finding Requirements Debt

The biggest challenge with type 1, poor requirements engineering practice, is that finding these problems is not as simple as running an automated tool. In many cases, there is a lack of useful information that is easy and ready to analyze. This is especially true in non-functional requirements (see box 3.2). Consider a company that plans to build a web app that manages home IoT (Internet of Things) information. If that company immediately began with the functionality for such an app without getting a clear, upfront understanding as to constraints and legal requirements, they would be incurring requirements debt. Of course, the fact that a company has legal constraints does not mean the company must deal with them,

Box 3.2
Types of Requirements

A common division of requirements is into *functional* and *nonfunctional*. Nonfunctional requirements (NFRs) are the quality criteria for a system. They include *ilities* such as maintainability, usability, availability, as well as performance and security. It is important to remember that although we call these nonfunctional, this is a poor term: they are often fairly easy to express in verifiable ways, and they often involve functionality that you need to implement. So we prefer the term "quality attribute" requirements. Quality attribute scenarios (QAS) are end-to-end requirements for, and acceptance tests of, a software system. These scenarios specify (at least) a *stimulus*, a *response*, and a *response measure*. For example, "when request is initiated the system responds within ten milliseconds."

as there are numerous reasons a company might ignore or be unaware of those requirements. Uber, for example, rolled out its ridesharing platform first and sought the required regulatory approval after it was successful. However, Uber still had those obligations, and arguably ignoring them was high risk. The question then becomes how a company can understand it might be taking shortcuts in the requirements process.

In this instance and others, the first sign of trouble might be a visit from a regulator or complaints and campaigns from annoyed users. Clearly this is far too late. In happier scenarios, the company may call in a consultant to advise on legal requirements, or a project staff member might be informed by a colleague. The best way to avoid the first type of problem is to have a solid requirements process. That isn't to say it has to be heavy-weight and onerous. But there should be (1) a well-known process by which requirements are gathered and prioritized and (2) a tool for managing requirements and feature requests. The second is the easiest to achieve. Many social coding tools provide an issue tracker that can easily handle the collection of requirements (and their traceability to features).[2] The first one seems to be more challenging. Particularly in start-ups, the time pressure to deliver features often means it is hard to know where new features and requests are coming from. This lack of visibility means it is fairly easy to lose track of why things are being delivered and what requirements have top priority. Is it because that requirement comes from the biggest customer? Or was it a lingering issue from a previous, outdated product? Or was it just the favorite requirement of the project manager or architect?

A well-defined process and tool also help with reducing our second type of requirements debt (ignorance of requirements). Modern approaches to software development, as characterized by a lean approach to software product delivery, focus on the *value stream*. This is the flow of useful, valuable parts of the software product from initial idea into a validated production setting. A value stream is only useful if we can quantify the value at any point in the stream. This implies tracking the feature throughout the software development lifecycle.

For example, if our hypothetical IoT web app wants to deliver a tool for monitoring a home's smart devices, we need to know where that feature came from (who requested it, who characterized it, who prioritized it), who is implementing it, and what testable implementation will give feedback on the success of that feature (e.g., number of new users).

3.2 Managing Requirements Debt

Managing requirements debt is about making the business case for doing good requirements elicitation and being aware of what requirements we are working on (and why). It means arguing why we should pay attention to the requirements process we are using and what requirements we are working on.

The value stream approach (figure 3.1; see also the Further Reading section for examples) is the easiest way to do this. A question that might be useful to trigger this type of analysis is: Why are we working on this feature? If no one can quickly and easily answer that question, you probably have requirements debt. Bonus points if the answer includes a verifiable, evidence-based reason why it is being worked on (e.g., "we expect this to reduce operator mistakes by 10%").

If this is not an easy question to answer, the first step is to analyze the tools that support your software development approach. Begin with the commits that were recently made to production. Can you tell why these commits were made? What issue do they fix or close? Ideally, every commit is either fixing a bug, paying down code and design debt (e.g., refactoring), or delivering a feature (see box 3.3).

If commits are traceable to an issue, debt, or feature, then begin looking at the issue tracker and backlog. What is sitting there? Can you tell where each issue came from? How long does it take an issue to move from being assigned, to being closed with a testable release? Is this time-to-close going up or down?

Requirements debt is higher if the time it takes to identify new features is longer than some threshold value. But what value? You can set this threshold by looking at your competitors or by polling your stakeholders. Keep in mind that, in some cases, changes can be too frequent. But if customer complaints and bugs sit too long on the backlog before being pulled into

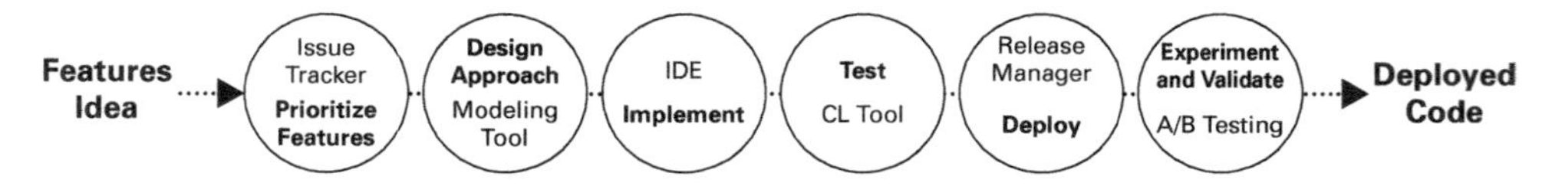

Figure 3.1
A value stream model, with sample tools. The arrows show how the idea moves from an issue tracker all the way to an A/B test result. This visualization can highlight bottlenecks, where a feature takes the most time to progress.

development and addressed, this is a sign of possible requirements debt. Again, working on the wrong thing is okay, if (and only if) it is not possible to understand its "degree of wrongness" until it is put into production. Ignoring obvious constraints—particularly legal and regulatory requirements—is inexcusable. The best way to manage and reduce requirements debt is to have a clear, tool-supported process for understanding the value of the work your team is doing.

Box 3.3

Requirements Traceability in Chromium

Requirements traceability is the ability to link specific, individual requirements to other software artifacts, such as code, design elements, and tests. Traceability ranges from formal traceability mechanisms using extensive traceability matrices to less formal automatically created links in IDEs and collaborative software platforms like Github. In the formal case, traceability links exist to satisfy regulators and to provide discoverability in the event of serious safety problems. In the informal case, traceability is more like simple linking (like when you tag a commit with an issue ID).

The widely used Chromium project, which includes the Chromium browser (the basis for Chrome, the most popular browser in the world, at the time of writing) and the Chromium OS, takes a rigorous approach to requirements traceability. All bugs are stored in the project's issue tracker (https://bugs.chromium.org). That does not distinguish the project. However, what does distinguish Chromium is the discipline with which bugs/issues are labeled, and the discipline with which commits are linked to bugs/issues. For example, bugs are labelled with bug types. You can search for type=bug-security or type=bug-regression to find just security bugs or just regressions. Bugs that are to be fixed are given a priority. Furthermore, when bugs are fixed (by commits) these commits reference the issue that they are fixing. This seems like a simple, even obvious, practice. But so many projects fail to do this. In many projects that we have seen bugs are reported and feature-requests are made and captured in an issue-tracker, and at some point commits are made, but they are unrelated; that is, there is no discipline of tagging, of saying "*this* commit is to resolve *that* issue." And without this discipline, it is very difficult to understand your project—to understand which parts of the code are related to the most critical bugs, for example, or to spot trends in the way that the project is evolving. We will return to this topic in chapter 4, but suffice it to say that a discipline of requirements traceability, like they do in the Chromium project, is going to be one of the foundations in a disciplined approach to monitoring and managing technical debt.

3.3 Avoiding Requirements Debt

To avoid accumulating requirements debt in the first place, we have two recommendations. Do a good job of requirements elicitation, and leverage crowd-sourcing techniques for gathering requirements.

3.3.1 Requirements Elicitation

Requirements elicitation is not innate. It is a skill that needs to be trained for and consciously developed in an organization. It usually involves some amount of marketing knowledge, domain expertise, and communications skills. Often people with this combination of skills are not developers. A common person to manage this might be a product (or project) manager (PM), someone with enough technical skill to talk with both developers and customers. Elicitation has several facets. One is to understand where to find requirements. In value-stream driven organizations, at least some of the features will come from production analytics. What are people clicking on? How long does it take to find a certain page? What features didn't get traction? PMs and other stakeholders (customer support, customer success manager) should work together on the requirements, crossing several data points (e.g., requirements from new customers, surveys from existing customers, etc.). A few hours spent on requirements elicitation today can save you hours of rearchitecture and implementation tomorrow (not counting time to fix bugs related to these potentially useless features).

Another facet of elicitation is to properly collect and manage the gathered requirements. This means understanding which requirements are similar, which ones are dependent on others, and which requirements are already implemented. This does not have to involve requirements management tools like IBM's DOORS, but tools do help to make things concrete. In highly regulated environments, such as automotive or aeronautics, creating an explicit model of requirements in a tool is itself likely to be a requirement mandated by standards (such as ISO26262 or DO-178C).

The final task of elicitation is to translate the requirements into manageable, nonambiguous, developer-understood language. This means three things:

1. Write the requirement in a form that developers and team leads can understand.
2. Filter out requirements that will never be implemented.
3. Phrase and add the requirements into the developer-facing tooling.

Developers understand the need for requirements. However, the reason someone else handles elicitation is that it is wasteful to have a roomful of engineers debating what a thermostat really is, for example. This is better done by a product manager (PM) or requirements engineering specialist who can translate the requirement into the same language the system itself uses. That does not mean specifying the design—that should never be part of a requirements process—but it does imply that the product manager understands how and where the feature might be implemented.

The product manager should also filter out requirements that they already know will never be implemented. This might be understood from existing prioritization decisions affecting personnel and budgets, or it might be a violation of some obvious (to the organization) constraints or work in progress. This is a dangerous proposition, of course, as it endows the PM with vast powers.

Finally, the PM should be responsible for moving or tracing the requirements from the requirements tool into a tool the developers use frequently. For example, requirements might be managed in a requirements tool like DOORS, and developers may work with tickets/issues in an issue-tracking tool such as Jira. The PM should ensure there is traceability and continuity between these different tools.

3.3.2 New Approaches to Gathering Requirements

Our other suggestion for gathering good requirements information is to leverage user feedback. The obvious way to do this is to go ask the users, and it is surprising how infrequently this is done in most organizations. The first issue is to figure out who the users are. A number of different approaches can be leveraged to get their feedback, including surveys (preferably unbiased and representative of your user base), A/B testing, and observation.

An emerging approach for understanding user feedback is via data analytics. This comes in two flavors. One, monitor and analyze all interactions with your software. Two, use natural language processing (NLP) to extract feature ideas from large text repositories.

The first approach means that, to the extent possible, all interactions with your software should be recorded. This means log file analysis in web environments. In apps, many third-party libraries offer the ability to record interaction with the user interface. In on-premises software, you may have to add your own internal logging services.

This dataset is incredibly valuable at analyzing how people are using and responding to the software. Their response to the software—for example, the amount of time spent on a certain page—is of course ultimately about how they are responding to the originally identified requirements. That gives you two insights. One, whether the existing requirements are valuable. Two, and more importantly, the data may reveal new potential for features. Consider a log that shows 5% of visitors leave just before making the purchase. This should suggest some feature, or improvement, is missing to convert these 5% to purchasers.

Many forums exist now for online discussion of software. For product software, such as apps or websites, these might be the relevant app store or general discussion forums. For custom software, you might even create your own feedback forum. These repositories are potentially rich sources of customer feedback. For app stores, for example, things like star ratings and user comments can be mined and processed by natural language processing techniques. One approach might be to first filter out spam and trivial comments of fewer than ten words, parse the text for keywords, and then to apply sentiment analysis to understand how people are disposed to the product. Another approach is to use the same set of reviews to find suggestions for new features (for example, "This would have gotten five stars if only it had X").

3.3.3 Recording Requirements

Having captured requirements, they should be correctly and adequately recorded. **Correctly** simply means the requirement reflects the information elicited. **Adequately** is context-specific. In a simple system, adequate might just be a list on a piece of paper, or as simple as saying "like the last one we did, but in blue" (e.g., for a company that creates restaurant websites). In more complex domains, or in regulated domains (such as avionics), adequate implies more detail.

For example, at large software companies like Microsoft, each requirement is defined in a ticket, and the ticket records, among other things:

- Requester
- Assignee
- Related components of existing software products
- Priority
- Related requirements

At this point, a tool is almost certainly essential. This might be as basic as a spreadsheet, but requirements-specific tools will provide additional

Box 3.4
Lean and Agile and Requirements Debt

There used to be a specific and well-known way to do requirements engineering. In older software processes, a separate phase was dedicated to requirements gathering and analysis, predating later waterfall phases in which the requirements would be concretely realized in a design and then implementation. Standards such as IEEE 29148 give useful and accurate definitions for specifying these types of requirements.

The only problem with this approach is that it turns out not to work too well for real-world software. Beginning with the agile manifesto (actually, most so-called waterfall theorists understood the need to be iterative and incrementally deliver a product), software engineers began to insist that a separate, up-front requirements phase was counterproductive. Instead, they called for a focus on working software. This was later expanded with the lean manufacturing focus introduced by people like Donald Reinertsen, and the book *The Goal* by Eliahu Goldratt. Modern thinking instead sees a process of constant change and feedback, where requirements are rapidly formulated into features, with a minimum of reflection and with a minimum of delay between the requirements being conceived and the user getting to experience the product.

One example of this is the Lean Startup philosophy described by Eric Ries. He suggests a requirement on its own has no value; only a requirement that can be tested in production helps the enterprise or organization learn and understand what the users really wanted.

In this context, one might ask whether it is possible to acquire requirements debt at all! After all, if requirements debt is caused by shortcuts in requirements gathering, analysis, and specification, then presumably a rapid feedback, frequent iteration model should drastically reduce this. If you prioritize the wrong requirement but find that out quickly, maybe the debt cost was not significant. Unfortunately one can still acquire requirements debt!

In more lean approaches, the debt occurs when we focus too much in any single iteration on building features and not enough on whether those are features users want. In particular, agile and lean approaches can suffer from short-term prioritization that ignores longer-term product goals. And the most likely source of debt is that the organization has no good way of linking the original requirement—or product idea—to the eventual release or feature toggle that instantiates it. As a result, it becomes very difficult to learn the value of a given requirement.

features, including the ability to link to other artifacts, such as design choices and code commits. Just like with deployment, the industry trend is toward increasing automation of requirements engineering, and tool support is key to this.

3.4 Summary

A low-debt requirements process is one that uses tooling to trace requirements all the way into production. It is one where it is easy to answer the question: Why are we working on this feature? The best way to avoid requirements debt is to have a good sense for what the process is and why it exists. It means having someone, even part-time, explicitly responsible for eliciting requirements and determining what to build and when.

Ultimately a low-debt organization is one where requirements flow, and flow quickly, into production with actionable feedback at the end (see box 3.4 on the lean approach). This implies that requirements have traceability all the way from elicitation into production. Ideally, one should be able to link a requirement to the particular build and experiment that implemented that requirement and to the data that argues for the importance of the requirement.

Notes

1. The full text of this and other Voice of the Practitioner sections can be found in the Appendix under the relevant name.

2. Including Github, Bitbucket, Gitlab.

Further Reading

The concept of lean and value stream mapping has taken hold of modern software practice. The notion initially came from automotive manufacturing and the Toyota Way. It was popularized in software engineering in David Anderson and Donald Reinertsen, *Kanban: Successful Evolutionary Change for Your Technology Business* (Blue Hole Press, 2010); as well as the writings of Mary and Tom Poppendieck, such as *Lean Software Development: An Agile Toolkit* (Addison-Wesley, 2003). Most recently, Eric Ries's book, *The Lean Startup: How Today's Entrepreneurs Use Continuous Innovation to Create Radically Successful Businesses* (Currency, 2011); and Mik Kersten, *Project to Product: How to Survive and Thrive in the Age of Digital Disruption with the Flow Framework* (IT Revolution Press, 2018) have argued for a focus not on shipping features but instead on delivering value that is easily evaluated in a process that permits radical, iterative release of working software.

Defining requirements has been done many times, but the definition we use is taken from I. J. Jureta, J. Mylopoulos, and S. Faulkner, S. "Revisiting the Core Ontology and Problem in Requirements Engineering" in *Proceedings of the IEEE Joint International Conference on Requirements Engineering*, 2008, 71–80.

Traceability was initially defined by Olly Gotel in "An Analysis of the Requirements Traceability Problem," in the *IEEE International Requirements Engineering Conference*, 1994. More recently, Jane Cleland-Huang and her students have focused on applying natural language processing approaches to automate requirements traceability, for example, J. Cleland-Huang, B. Berenbach, S. Clark, R. Settimi, and E. Romanova, "Best Practices for Automated Traceability," *Computer* 40, no. 6 (June 2007): 27–35.

For natural language processing of app stores, a good place to start is the work of Walid Maalej, Zijad Kurtanovic, Hadeer Nabil, and Christoph Stanik, "On the Automatic Classification of App Reviews," *Requirements Engineering* 21, no. 3 (2016): 311–331.

The most commonly used standard for requirements specification is the Institute of Electrical and Electronics Engineers (IEEE) standard 29148, which is available at https://ieeexplore.ieee.org/document/8559686. A lighter-weight standard is EARS, the Easy Approach To Requirements Syntax at https://ieeexplore.ieee.org/document/8559686.

4 Design and Architecture Debt

—with Yuanfang Cai

> Art is to be free. Design is to fix.
> —Kanye West

> Successful design is not the achievement of perfection but the minimization and accommodation of imperfection.
> —Henry Petroski

In this chapter we turn our attention to design. Unlike code-based technical debt, which we discuss in the next chapter, design debt is hard to spot because it is more diffuse: its causes are distributed across a system's files and are often only detectable when analyzing a project's history and its trends. We are going to assume we understand what to build—that is, we have little requirements debt. The main message of this chapter is that design debt is highly critical to a project's long-term success, and that it can in fact be detected and remedied. The way technical debt manifests itself in design or architecture is similar, so we will use the term "design debt" to refer to both design and architecture debt.

Technical debt is often caused by poor choices that are made for reasons of expediency. Mature software development instead consciously creates coding standards and design rules or guidelines to shape the evolution of the project and provide conceptual integrity. A key skill of lead developers, team leads, and software architects is the ability to create effective design rules. A design rule is an organization-wide principle that defines how software design questions should be resolved or managed, and in particular, which quality requirements are most important. Design rules can have profound effects on project quality and productivity. An architect might create rules regarding layering, the pervasive use of patterns such as model-view-controller (MVC), model-view-presenter (MVP) or abstract factory, or might mandate that certain frameworks are only invoked through an intermediate interface to hide implementation details. Such rules help guide the evolution of the project and are intended to provide conceptual integrity.

But designs are subject to the laws of entropy, as are all software artifacts. Without careful attention and the input of energy, design rules are undermined, the impact of those rules decreases, and the design itself loses conceptual integrity. We call this particular form of entropy design debt, a specific kind of technical debt. Design debt is a particularly pernicious form of technical debt, since by definition it involves nonlocal concerns. A nonlocal concern is an aspect of the software that touches on multiple classes, files, or modules, like security or performance. Since the concern is not confined to a single piece of code, it is not easily addressable by debt detection and removal tools, code-level metrics, code inspections, and so on. For example, a relatively common form of debt is when there are cyclic dependencies among a group of files. Because this cycle of dependencies may go through several files, it is easy to miss; you need to be intimately familiar with *all* of the files in the cycle to be aware of this debt instance. In large systems, this might mean that some of the files are not even within the control of your development team.

Thus the lead developer or architect's job is not just making design rules but also shepherding the project and its people. In chapter 10 we touch more on these social aspects of debt. The lead developer's job is to ensure that the rules are properly followed throughout the project's lifetime, through all of the bug-fixing and feature-adding activities of developers, so that the integrity of the design is maintained. In the case where the architecture has been allowed to degrade—perhaps in the rush toward a deadline—the architect's more critical tasks are to find the sources of debt, to quantify their impact, and to refactor the ones that present a risk to the long-term health of the project. Architectural debt is costly. One study found that it was, in fact, the leading source of technical debt.

Happily, as we will see, with design debt it is relatively straightforward to calculate the return on investment (ROI) of refactoring to remove debt. It is rather rare in software engineering to be able to make a true business case for implementing best practices, but in this case we really can do it, as we will show in section 4.2: Managing Design Debt.

4.1 Identifying Design Debt

There are several causes of design debt, for example:

- Improper separation of concerns—where multiple responsibilities are implemented in the same class, making the class bloated, difficult to understand, test, and modify.
- Clone and own—where a piece of code is copied (cloned) and then modified, typically because the programmer does not understand the dependencies on the original code and so fears to modify it or take time to understand it.
- Tangled dependencies—sometimes called "big ball of mud," where there has been little thought given to architectural issues and so design activities proceed implicitly.
- Unplanned evolution—where new features and bug-fixes are added without any consideration of how these changes affect the overall structure and conceptual integrity of the system. This is rather common today in most teams where attention is devoted to fixing the immediate, pressing issue—implementing a feature or fixing a bug—without paying much attention to the impact of such changes in terms of maintainability, readability, or modifiability.

This list is not exhaustive, but these are some of the most common causes of design debt. You might ask: Why do these practices lead to debt, and how do we know this? Quite simply, they lead to debt because they violate the principles of good design: these practices tend to increase coupling or decrease cohesion (see box 4.1). They degrade the modular structure of a system. Note that by "module" here we are referring to a unit of development and maintenance. Depending on the language, a module could be a package (as in Java or Python) containing classes, or a group of files in a directory (as in C), containing files. Other realizations of modules are also possible.

Consider the five SOLID principles for object-oriented design:

1. *Single responsibility principle*: every class or module should have just a single responsibility, and that responsibility should be encapsulated by the class or module.
2. *Open/closed principle*: "software entities . . . should be open for extension, but closed for modification," meaning that a class can be extended, for example through inheritance, without modifying the original entity.

Box 4.1
Coupling and Cohesion: A Primer

The concepts of coupling and cohesion were first introduced to the software engineering community by Larry Constantine in the late 1960s. Essentially coupling, as its name suggests, refers to the degree to which two modules depend on each other. This dependency might be purely structural—for example, code in one module calling a function in another module—or it might be data related, such as having two modules both depend on the structure of a file or database or message format. In addition, there are other types of coupling. For example, two modules might be logically coupled—module A needs to complete a correct execution before module B can execute—even if there is no detectable dependency between A and B. Or they may be temporally coupled: for example, modules A and B are bundled together into a single process because they need to execute at the same time. Typically we strive for low coupling in software development as that allows modules to be more easily and independently developed, debugged, and modified.

Low coupling tends to be associated with high cohesion. Cohesion is the degree to which the responsibilities of a module are related to each other. For example, if you have pricing logic and only pricing logic in one module of your e-commerce website, that module would be highly cohesive. But if that module also included logic to format prices in HTML forms and to store and retrieve prices from a database, that module would be less cohesive, as it now includes responsibilities that are not directly related to pricing logic. Typically we strive for high cohesion in software development as that allows a programmer to know where to look and where to change if there is a bug or a change in requirements.

3. *Liskov substitution principle*: objects should be replaceable with instances of their subtypes without altering the correctness of a program. That is, it should always be possible for you to replace a class with one of its subclasses without affecting the behavior or properties of the program.
4. *Interface segregation principle*: using several specific interfaces is better than one general-purpose interface. The typical explanation of this principle is that no client should be forced to depend on interfaces that it does not use. This says that interfaces should be as specific as possible, which is, in essence, the same as saying that they should be cohesive.
5. *Dependency inversion principle*: "High-level modules should not depend on low-level modules. Both should depend on abstractions," and that

"abstractions should not depend on details. Details should depend on abstractions." The idea here is that lower-level modules should provide interfaces (which they implement) to higher-level modules. This decouples them and allows the lower-level implementation to change without affecting clients of the interface.

The SOLID principles have been codified for over three decades. These principles, and others like them (for example, the general responsibility assignment software patterns, known as GRASP, and see Box 4.2), describe ways of structuring software so that it has high cohesion and low coupling, thus increasing modularity and hence the ability of changing, combining, and replacing modules in a system. Creating software that follows these principles takes more time up front, for sure. It takes more time, for example, to create an abstract interface between two layers in your software than simply having the upper layer make calls to the lower one. But the benefit of doing so is that when the lower layer's implementation changes—but its interface stays stable—the upper layer does not need to be modified.

Thus the architect or developer has a pay-me-now-or-pay-me-later choice to make with every class that they work on. Of course you would not create an abstraction everywhere; we need to be mindful of extreme programming's YAGNI (you aren't gonna need it) dictum. But if you know, and if experience tells you, that the dependencies between two major subsystems or layers are going to change, then you likely *will* need it—the abstraction that is!

Violating any of these principles invariably results in design debt, because violating them leads to situations where changes cannot be made in a clean way, since responsibilities are needlessly tangled together. As we have said, not all debt is bad debt. Sometimes a principle is violated when there is a worthy tradeoff. For example, it is common to sacrifice low coupling or high cohesion to improve performance or time to market. Thus the cost and impact of each violation need to be measured.

How to decide what to do? As we said in chapter 1, if you cannot measure design debt (or any other kind of debt) then you cannot manage it. So we will now discuss some of the consequences of design debt. We will show how to measure it and make principled decisions about if, when, and how to pay down the debt.

Box 4.2
Design Principles

Our focus in this chapter on object-oriented (OO) principles like SOLID reflects our familiarity with OO more than any special characteristics of OO approaches to design. Functional programming, for example, has principles for good design that are often enforced by the language semantics itself: immutability, compositionality, and referential transparency are language elements in Scala or Haskell that force programmers to write cleaner code. However, concepts such as information hiding and separation of concerns are still highly important to these programs.

There are other, paradigm-specific language features as well, such as the principles of reactive web development, actor models of concurrent programming, or effective database design. In each paradigm there are design rules that help maintain the conceptual integrity of the system but can and will degrade with time if not invested in.

4.1.1 Quantifying Design Debt

In any real debt in life, we would like to know the principal and the interest rate on this debt. This is just common sense, and we see this information plainly printed on every mortgage, credit card, or car loan statement. But in software development both the principal and interest are seldom known and seldom calculated. Their true values are obscured by the messy day-to-day details of what we perceive to be our main job—implementing features and fixing bugs. Most developers and architects have only a vague notion of the accumulated debt in their project. They sense the debt, and they feel it and struggle with it in their daily activities, but they cannot say what its true value or impact is, or precisely where it lies. So how can we quantify this debt?

To quantify the debt, we first need a concise definition of design debt. A design debt is: (1) a group of connected modules (typically files)—the principal, and (2) a model of their maintenance costs and how these grow over time—the interest. Let's examine this definition in detail.

First, how do we determine if a group of files are connected? There are two relatively simple ways to do this. The first is to determine the static dependencies between the files in your project. You can find these by employing a static code analysis tool such as Understand, which can extract the dependencies between files (for example, calling dependencies or inheritance dependencies). A dependency might exist, for example, when some code in one file calls a method in a second file. The second is by capturing the evolutionary dependencies between files in your project. An evolutionary dependency occurs when two files change together, and you can get this information from your revision control system. The reason we need to capture both kinds of dependencies is that either one, by itself, does not give a full picture of the dependencies that you care about in a software system. Two files may change together and yet not have any structural dependency—this occurs when, for example, these two files share some assumption or knowledge. For example, two files may assume the same organization of data in a file or database, or they may assume the same communication protocol, message format, key size, or metadata, and so on.

Consider the example shown in figure 4.1. In this example, file dependencies are represented using a design structure matrix, or DSM. DSMs have been used in engineering design for decades and are currently supported by a number of industrial tools such as Lattix,[1] Silverthread,[2] and DV8.[3] In a DSM entities of interest (in our case, files) are placed on the rows of the matrix and also, in the same order, on the columns. The cells of the matrix are annotated to indicate the type of dependency. For example, we can annotate a DSM cell with information showing that the file on the row inherits from the file on the column, or that it calls the file on the column, or that it co-changes with the file on the column. The first two annotations are structural and the third is an evolutionary (or history) dependency. Of course, the cells on the diagonal represent self-dependency.

The example shown in figure 4.1 is a small excerpt from an industrial project, showing twenty-seven of its files and their dependencies. (The actual file names have been changed, but that does not affect the point of this example.) For example, this figure shows that the file on row 4—FederatedLoginController.java—creates (an instance of) the file on column 1—MenuBean.java—and that the file on row 18—ReportGenerationController.java—uses the file on column 11—BaseGenerationController.java. These static dependencies are easily extracted by reverse-engineering the source code (or, in many cases, compiled code).

As you can see from a cursory examination of figure 4.1, this matrix is quite sparse. This is normally a good thing! It means that the files are not heavily coupled to each other and, as a consequence, you might expect that it would be relatively easy to change these files independently. In other

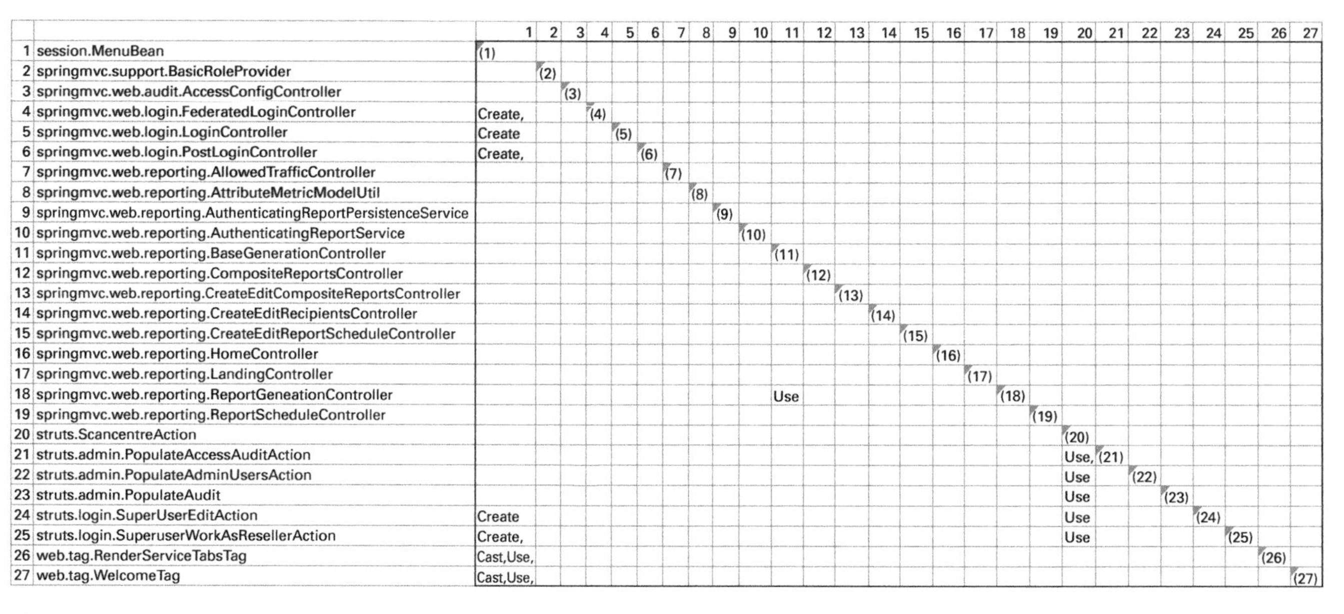

		1	2	3	4	5	6	7	8	9	10	11	12	13	14	15	16	17	18	19	20	21	22	23	24	25	26	27
1	session.MenuBean	(1)																										
2	springmvc.support.BasicRoleProvider		(2)																									
3	springmvc.web.audit.AccessConfigController			(3)																								
4	springmvc.web.login.FederatedLoginController	Create,			(4)																							
5	springmvc.web.login.LoginController	Create				(5)																						
6	springmvc.web.login.PostLoginController	Create,					(6)																					
7	springmvc.web.reporting.AllowedTrafficController							(7)																				
8	springmvc.web.reporting.AttributeMetricModelUtil								(8)																			
9	springmvc.web.reporting.AuthenticatingReportPersistenceService									(9)																		
10	springmvc.web.reporting.AuthenticatingReportService										(10)																	
11	springmvc.web.reporting.BaseGenerationController											(11)																
12	springmvc.web.reporting.CompositeReportsController												(12)															
13	springmvc.web.reporting.CreateEditCompositeReportsController													(13)														
14	springmvc.web.reporting.CreateEditRecipientsController														(14)													
15	springmvc.web.reporting.CreateEditReportScheduleController															(15)												
16	springmvc.web.reporting.HomeController																(16)											
17	springmvc.web.reporting.LandingController																	(17)										
18	springmvc.web.reporting.ReportGeneationController											Use							(18)									
19	springmvc.web.reporting.ReportScheduleController																			(19)								
20	struts.ScancentreAction																				(20)							
21	struts.admin.PopulateAccessAuditAction																				Use,	(21)						
22	struts.admin.PopulateAdminUsersAction																				Use		(22)					
23	struts.admin.PopulateAudit																				Use			(23)				
24	struts.login.SuperUserEditAction	Create																			Use				(24)			
25	struts.login.SuperuserWorkAsResellerAction	Create,																			Use					(25)		
26	web.tag.RenderServiceTabsTag	Cast,Use,																									(26)	
27	web.tag.WelcomeTag	Cast,Use,																										(27)

Figure 4.1
A DSM of an industrial project showing structural dependencies.

words, it looks like there is relatively little design debt here. Sadly, this is not the case. In figure 4.2, we add in evolutionary dependency information as numbers in the cells. For example, the file on row 20 changed sixteen times with the file in column 1, as extracted from the project's commit history.

A quick glance at figure 4.2 gives us a very different picture of the project. The matrix is now rather dense. (The details of this matrix are not what is important here. What you should take away is that this matrix has *many* more cells populated.) Although these files are, by and large, not structurally coupled to each other, they are strongly evolutionarily coupled! Furthermore, we see many annotations in cells above the diagonal in the matrix. This means that the coupling is not just from higher-level to lower-level files, but it goes in all directions. This project, in fact, suffers from very high architectural debt, as reported by the architects—almost every change in the project is costly and complex, and predicting when new features will be ready or when bugs will be fixed is challenging.

While this kind of general insight is interesting, we can do better: we can precisely quantify the costs and impact of the debt.

Assuming that you have captured these two kinds of dependencies for your project, you can calculate a growth model of maintenance costs. For any set of connected, interdependent files you can capture their historic maintenance costs as the amount of churn (committed lines of code) spent on fixing bugs, as of a specified time period (e.g., the end of a release). And you can track, for each set of files, how this cost has grown over time. You can very easily compare these growth rates with project averages to determine which groups are likely to contain debt. We will discuss precisely how to do this shortly.

4.1.2 Gathering Relevant Information

There are three kinds of information—that virtually every project collects or can obtain—that provide insight into design debt. These are:

- Issues, from an issue-tracking system such as Jira (or tickets from platforms such as GitHub or GitLab)
- Commits, from a revision control system such as Git (or Subversion)
- Source code in one or more programming languages

By reverse-engineering the code, we can determine the (static) structural relationships between code modules. If module A inherits from module B, a

		1	2	3	4	5	6	7	8	9	10	11	12	13	14	15	16	17	18	19	20	21	22	23	24	25	26	27
1	session.MenuBean	(1)		,10		,10										,8	,8				,16				,8			
2	springmvc.support.BasicRoleProvider		(2)													,8												
3	springmvc.web.audit.AccessConfigController	,10		(3)				,8					,8	,8		,10	,10	,8		,8	,8							
4	springmvc.web.login.FederatedLoginController	Create,			(4)	,10	,8									,8	,8				,10							
5	springmvc.web.login.LoginController	Create,10			,10	(5)															,14							
6	springmvc.web.login.PostLoginController	Create,			,8		(6)																					
7	springmvc.web.reporting.AllowedTrafficController			,8				(7)		,8			,12	,16	,14	,16	,10	,10	,8	,10								
8	springmvc.web.reporting.AttributeMetricModelUtil								(8)								,10											
9	springmvc.web.reporting.AuthenticatingReportPersistenceService							,8		(9)	,8		,10	,10		,12	,8	,10	,12									
10	springmvc.web.reporting.AuthenticatingReportService									,8	(10)					,10	,8											
11	springmvc.web.reporting.BaseGenerationController											(11)				,10			,8									
12	springmvc.web.reporting.CompositeReportsController			,8				,12		,10			(12)	,14	,10	,12	,10	,10	,10	,8								
13	springmvc.web.reporting.CreateEditCompositeReportsController			,8				,16		,10			,14	(13)	,14	,16	,10	,10	,10	,10								
14	springmvc.web.reporting.CreateEditRecipientsController							,14					,10	,14	(14)	,14	,10	,8		,10								
15	springmvc.web.reporting.CreateEditReportScheduleController	,8	,8	,10	,8			,16		,12	,10	,10	,12	,16	,14	(15)	,14	,14	,10	,14	,10							
16	springmvc.web.reporting.HomeController	,8		,10	,8			,10	,10	,8	,8		,10	,10	,10	,14	(16)	,12		,8	,10					,8		
17	springmvc.web.reporting.LandingController			,8				,10		,10			,10	,10	,8	,14	,12	(17)										
18	springmvc.web.reporting.ReportGeneationController							,8		,12		Use,8	,10	,10		,10			(18)									
19	springmvc.web.reporting.ReportScheduleController			,8				,10					,8	,10	,10	,14	,8			(19)								
20	struts.ScancentreAction	,16		,8	,10	,14										,10	,10				(20)		,8	,10	,8	,8		
21	struts.admin.PopulateAccessAuditAction																				Use,	(21)		,14				
22	struts.admin.PopulateAdminUsersAction																				Use,8		(22)					
23	struts.admin.PopulateAudit																				Use,10	,14		(23)				
24	struts.login.SuperUserEditAction	Create,8																			Use,8				(24)	,10		
25	struts.login.SuperuserWorkAsResellerAction	Create,																			Use,8				,10	(25)		
26	web.tag.RenderServiceTabsTag	Cast,Use,																									(26)	,8
27	web.tag.WelcomeTag	Cast,Use,																									,8	(27)

Figure 4.2

A DSM of the project in figure 4.1, now showing evolutionary dependencies.

parser can find this dependency information and we can store it in a representation such as the one shown in figure 4.1. Dynamic relationships are a bit more tricky. Dynamic relations are found in dynamically typed programming languages such as Python, Javascript, and Ruby. Dynamic relations are also found in distributed systems employing message-passing. In these cases you cannot reliably know, via a static code analysis, what architectural relations exist. In these cases you typically need to insert some form of instrumentation or monitoring and then run the instrumented version of your system to collect such data.

In addition, using the project's commits we can determine historical co-change relationships, as described above: for each commit we record the pairs of files (if any) that changed together, and we keep running totals of these co-commits as illustrated in figure 4.2. Assuming the project follows the approach in chapter 3, using a reasonable discipline to annotate every commit with the issue ID, from the issue repository, then we can know the reason for this commit. Such a discipline can be enforced in code reviews. See chapter 5 for more on this point. We can determine, for example, which commits were made to fix bugs and which commits were made to implement new features. By connecting issue information with revision history, we can easily keep track of the amount of churn—committed lines of code—to fix those bugs and implement those features.

All of this can be done trivially with information that most projects already collect, but which they do not currently analyze. If your project does not have the practice of recording an issue ID with every commit, it should do so! Without this information it is challenging to know why any change was made.

4.1.3 Analyzing the Gathered Information

Once at least some of this dependency information and history information has been gathered, you can do some useful analyses. It is very interesting to know what files are clustered together due to dependencies. This information is valuable because dependencies are a form of coupling, and high coupling between files means that it is difficult to change any one file without impacting others. Coupling might be explicit (such as when a method in one class calls a method in another class) or implicit (such as when two classes share some assumption). These implicit assumptions can vary widely: perhaps a set of files implicitly share knowledge of a data transmission protocol, a key length, the ordering of events, or knowledge about how system data values can change depending on the state or mode of the system. In such cases if one file is changed—perhaps the data transmission format is altered—then the others very likely will need to be changed as well.

If you have to pay attention to every change that you make, and a change may mean modifying multiple source files, this increases the cost of each change, whether that change is made to implement a feature or to fix a bug. This is a form of debt. And this additional cost, it must be stressed, is often incidental, and not inherent. That is to say, some complexity is inherent in your code base—perhaps you have a very complex set of business rules. But some complexity is due to the interest that has accumulated over time as various developers have made changes, each of which is done in isolation.

The picture in figure 4.3 shows some real-world examples of this kind of accumulation of design complexity. In these cases we can easily see the

Figure 4.3
Design complexity and design flaws in the real world.
Source: left image: https://pxhere.com/en/photo/996703; right image: ©Rick Kazman.

design complexity—it is immediately obvious. But in software the complexity accumulates invisibly. We can categorize these sorts of mistakes as design flaws or architecture antipatterns. Let's look at a couple of examples of such antipatterns.

Files that are connected to each other in a cycle of dependencies (a method in class A calls class B which calls class C which calls class A) are highly coupled. This cyclic coupling can usually be broken by refactoring—moving some functionality from, for example, class C to class A. These kinds of cycles can be found in virtually all systems. Figure 4.4 shows an example from the Chromium project. Files 13–17 form a cluster, with almost all of the files depending on—using or calling—each other.

Another common kind of incidental coupling occurs in inheritance hierarchies. Consider, for example, the example shown in figure 4.5, reverse engineered from Apache Hadoop. This DSM shows the FileSystem class (org.apache.hadoop.fs.FileSystem)—an important abstraction in Hadoop, along with seven classes that inherit from FileSystem: FilterFileSystem, RawLocalFileSystem, S3FileSystem, KosmosFileSystem, DistributedFileSystem, HftpFileSystem, and RawInMemoryFileSystem. The inheritance relationship is indicated in the DSM by the "ih" notation. Note, however, that the FileSystem class depends on (via a calling relation) the DistributedFileSystem class, as indicated by the "dp" annotation in cell (1,6). This means that the parent class calls the child class, which is a violation of good

		1	2	3	4	5	6	7	8	9	10	11	12	13	14	15	16	17	18	19
1	chrome.renderer.render_view.h	(1)	Use																	
2	chrome.renderer.render_thread.h		(2)																	
3	chrome.browser.extensions.extension_message_service.cc			(3)																
4	chrome.renderer.extensions.renderer_extension_bindings.h	Use			(4)															
5	chrome.browser.debugger.extension_ports_remote_service.cc			Use		(5)	Use													
6	chrome.browser.debugger.extension_ports_remote_service.h						(6)													
7	chrome.renderer.extensions.event_binding.h	Use	Use					(7)												
8	chrome.browser.automation.extension_port_container.cc			Use					(8)	Use										
9	chrome.browser.automation.extension_port_container.h									(9)										
10	chrome.renderer.resources.event_binding.js										(10)									
11	chrome.renderer.resources.extension_process_binding.js										Call	(11)								
12	chrome.browser.extensions.extension_message_unittest.cc			Use									(12)					Call		
13	chrome.renderer.extensions.extension_process_binding.cc	Call												(13)	Call		Use	Use		
14	chrome.renderer.render_view.cc	Call,Use	Call											Call	(14)	Call	Call	Call		
15	chrome.renderer.render_thread.cc	Call,Use	Call,Use											Call	Call	(15)	Call	Call		
16	chrome.renderer.extensions.event_binding.cc													Call	Call	Call	(16)			
17	chrome.renderer.extensions.renderer_extension_bindings.cc																Call,Use	(17)		
18	chrome.renderer.extensions.extension_process_binding.h																		(18)	
19	chrome.common.render_messages_internal.h																			(19)

Figure 4.4

Example of a cyclic dependency.

		1	2	3	4	5	6	7	8
1	org.apache.hadoop.fs.FileSystem	(1)					dp,26		
2	org.apache.hadoop.fs.FilterFileSystem	ih ,5	(2)						
3	org.apache.hadoop.fs.RawLocalFileSystem	ih ,5	,5	(3)					
4	org.apache.hadoop.fs.s3.S3FileSystem	ih ,8	,4	,6	(4)				
5	org.apache.hadoop.fs.kfs.KosmosFileSystem	ih				(5)			
6	org.apache.hadoop.dfs.DistributedFileSystem	ih ,26	,6	,7		,9	(6)		
7	org.apache.hadoop.dfs.HftpFileSystem	ih						(7)	
8	org.apache.hadoop.fs.InMemoryFileSystem$RawInMemoryFileSystem	ih ,7	,5	,8	,7		,9		(8)

Figure 4.5
Architecture antipatterns in Apache Hadoop.

object-oriented design practice, as it violates the SOLID principle of Liskov substitution (which states that any class should be able to be replaced by one of its subclasses).

This unnecessary dependency adds cost to the maintenance effort. Note that the cells (1,6) and (6,1) are annotated with the number twenty-six. This represents the number of times that FileSystem and DistributedFileSystem were co-committed in changes, according to Hadoop's revision history. This excessively high number of co-changes is a form of design debt. It is not only just a debt because a change to FileSystem usually requires a change to DistributedFileSystem but also because someone has to know this and to remember this implicit dependency. Fortunately this debt can be easily removed (that is, paid down) via refactoring: moving some functionality from the child class up to the parent. Note also that there are suspiciously high numbers of co-change relationships among the child classes; we can see co-change numbers between four and nine in the cells showing the dependencies among the child classes. Child classes should be independent of each other and depend only on the parent class! This, again, suggests incidental coupling which would need to be removed by refactoring.

It must be stressed, once again, that this kind of analysis is available today, using commercial and research tools and employing information that you are already collecting as part of your normal development activities (or which, if you are not collecting, can be easily collected with some inexpensive tools and some small changes to your development processes). Numerous studies have shown that the majority of lines of code committed in a project can be attributed to files that participate in architectural flaws. Think about this for a moment: if you care about your project then you want to spend your time and effort implementing features (and, sadly, fixing bugs), but you actually end up spending your effort in paying, over and over, for the flaws that have crept into your code base.

So what are these flaws exactly? Fortunately there are just a handful of flaw types. Six common architecture antipatterns—recurring design flaw types—have been identified over the past decade:

1. *Unstable interface*: where an influential file changes frequently with its dependents as recorded in the revision history
2. *Modularity violation*: where structurally decoupled modules frequently change together
3. *Unhealthy inheritance*: where a base class depends on its subclasses or a client class depends on both the base class and one or more of its subclasses
4. *Cyclic dependency* or *clique*: where a group of files form a strongly connected graph
5. *Package cycle*: where two or more packages depend on each other (rather than forming a hierarchical structure, as they should)
6. *Crossing*: where a file with both high fan-in and high fan-out changes frequently with its dependents and the files it depends on

These antipatterns have been shown to be very strongly correlated with bugs, changes, and churn (effort). We will show some data in a moment, but for now just keep in mind that if a file is involved in more design flaws (for example, a single file might participate in a modularity violation with one set of files, a cyclic dependency with another set of files, and an improper inheritance with a third set of files) then it is *far* more likely to be buggy and to require more effort to fix or modify. So . . . design flaws matter! They do not matter in some abstract sense of the conceptual integrity or beauty or elegance of the design; they matter in real terms: the person-days of effort that go wasted in dealing with them and their consequences. In box 4.3 we provide a brief discussion of how design flaws matter for security.

4.2 Managing Design Debt

As stated above, you cannot manage what you cannot measure. That is why identifying and quantifying design debt is so important. Without this measurement and quantification, it is virtually impossible to make a business case for removing the debt.

Box 4.3
Finding Security Design Flaws in Chromium

To get a feel for the broader impact of design flaws, let us consider an example: the Chromium project. Chromium is a very large open-source project—a web-browser and operating system—with a long, well-documented history. This makes it ideal as a case study for analysis. For this example we just focus on security bugs in Chromium. When we examine these, an interesting picture emerges. What we did is to extract what we call "Plus10" bugs. These are bugs that were labelled as "Type=Bug-security" in Chromium's issue tracker and where the resolution of these bugs required commits to ten or more distinct source files. The assumption here is that if a change requires modification to ten or more files, this is not simply a low-level coding issue—there is likely a design problem that needed to be addressed. For example, it might be the case that some functionality was improperly distributed across code modules, or that modules were too highly coupled, or that some shared "secret" or assumption needed to be changed. And this resulted in many simultaneous changes. In each of these cases the problem, at its heart, is a design problem.

Although these Plus10 bugs accounted for just 6.7% of all security bugs in Chromium, fixing them required changes to 44.6% of all the files ever modified to fix any security bug in the entire history of the project. This shows that a small proportion of bugs can have a very large impact. In addition, this set of Plus10 bugs covered 47.1% of all the lines of code spent to fix all security bugs. On average, each of these bugs required about seven times more lines of code (and so, we assume, about seven times more effort) to fix than other security bugs. Finally, these bugs required more commits than other security bugs.

Perhaps you are still skeptical about this set of data. Perhaps it is the case that these are just large, complex problems, and big problems require lots of time and effort to fix. This would be a reasonable doubt to have. But now let us consider one more characteristic of these Plus10 bugs: on average, the set of files modified to resolve these bugs contain 7.37 architectural flaws. Now the complexity of these bugs starts to make sense.

Consider, for example, Chromium issue 58,069: "Windows Sandbox allows access to the console."[4] This security bug required four commits to thirteen different files before it was finally closed, and these commits included modifications to 4,682 lines of code. The files that were modified were architecturally connected and five architectural flaws are detected in the relationships among these files, including two unstable interfaces, one clique, and two modularity violations.

This pattern repeats itself over and over in the Chromium project. Almost every Plus10 bug suffers from architectural antipatterns, and most suffer from multiple instances of such antipatterns.

Chromium Bug 34,151: "ChromeFrame: cookie policy not honored in chrome Frame" is another example where a vulnerability became extremely hard to fix due to the complex architectural relations among the involved files. This bug concerns the ChromeFrame plugin that violated Internet Explorer's cookie policy. To fix this problem, the cookies should be read from the host (i.e., Internet Explorer) when the ChromeFrame renderer requests them. The fix for this issue involved more than twenty files containing multiple flaws. First, *chromeFrameActivexBase.h, chromeFrameNpapi.h, chromeFrameNpapi .cc, chromeFrameDelegate.h,* and *chrome-FrameDelegate.cc* all implemented the method *OnGetCookiesFromHost,* and there is no structural dependency among these modules. There is a lack of abstraction here that manifests itself as a modularity violation. Secondly, the method *OnResponseStarted* is modified to un-use persistent cookies in *pluginUrlRequest.h,* and this in turn affects five other files. This is an instance of an *unstable interface*. Third, a new method *IsExternal* in *chromeUrlRequest_context.h* affects *automationProfileImpl.cc* and *resourceMessageFilter.cc,* another unstable interface. These three design flaws ultimately resulted in the team writing 528 lines of code over twenty files to make the eventual fix.

Every nontrivial software project has technical debt, including design debt. At this point you should be asking yourself the following question: Which groups of files in my project are involved in the most costly architecture debt? Once you have answered this question, the next obvious question is: What is the ROI of paying down this debt? And once you have answered these two questions, the case for paying down the debt (typically through refactoring) is a simple one; it is an economic decision. If the ROI is large, do the refactoring. We will explain this process with a case study done with a commercial software project where we identified design debt, quantified this debt, and estimated the cost of refactoring to remove the debt. Based on the estimated benefits of removing the debt and the costs of the refactoring to remove the debt, we were able to arrive at an expected ROI value.

This case study was done with SoftServe, a multinational software outsourcing company. SoftServe has a reputation for being a forward-thinking and disciplined software engineering organization. They are a mature software company and have adopted many best practices in architecture, testing, agile development, process, and project management.

Prior to our case study—a web portal system that we named SS1—SoftServe architects had already compiled a list of technical debt items in the project. These debts were detected by various tools, such as code violations detected by SonarQube, and many Todo and FIXME tags reported by Eclipse. These tools flagged so many problems that the developers were overwhelmed; they could not decide what and where to fix. There were also other problems with the debts that these tools reported: there was no estimate of the magnitude of the debt incurred by each violation or the value that might be gained by paying down these debts. Our goal, in contrast, was not only to understand the nature and magnitude of the design debt in SS1 but also to quantify these debts, to prioritize the various problems, and to quantify the return on investment of paying down the debt.

At the time of the analysis SS1 contained 797 source files, and we captured its revision history and issues over a two-year period. SS1 was maintained by six full-time developers and many more occasional contributors. During the period that we studied there were 2,756 issues recorded in their Jira issue-tracker (1,079 of which were bugs) and 3,262 commits recorded in their Git version control repository. The analysis process that we followed is illustrated in figure 4.6. It consisted of collecting data, using a tool called DV8 to identify design debt, then validating that debt with the team. Finally, we were able to quantify the debt costs.

At the heart of this analysis pipeline was DV8, a tool for clustering, visualizing, manipulating, and analyzing architectural relations.[5] By enacting the process depicted in figure 4.6 a set of hotspots were identified. These were clusters of architecturally-related files that covered the most bug- and change-prone files in the project and that were identified as participating in design flaws and architecture antipatterns.

In the end, three hotspots—clusters of architecturally-related files, called *DRSpaces*—were identified as containing the most debt in the project. The term "DRSpace" means "design rule space"—a group of related files led by a design rule. As we mentioned at the start of this chapter, a design rule is an important abstraction or interface in the system. For example, if you employ an abstract factory pattern, the design rule is that the factory defines a generic interface that is then implemented by specific subclasses that are instantiated by clients of the abstract factory. The abstract factory interface leads a set of files that depend on this abstraction.

Each of the three DRSpaces identified in the SS1 project were similarly led by an important file, one that influenced many other files in the project. The leading file of each DRSpace is shown in figure 4.7, in cells A2, A3, and A4.

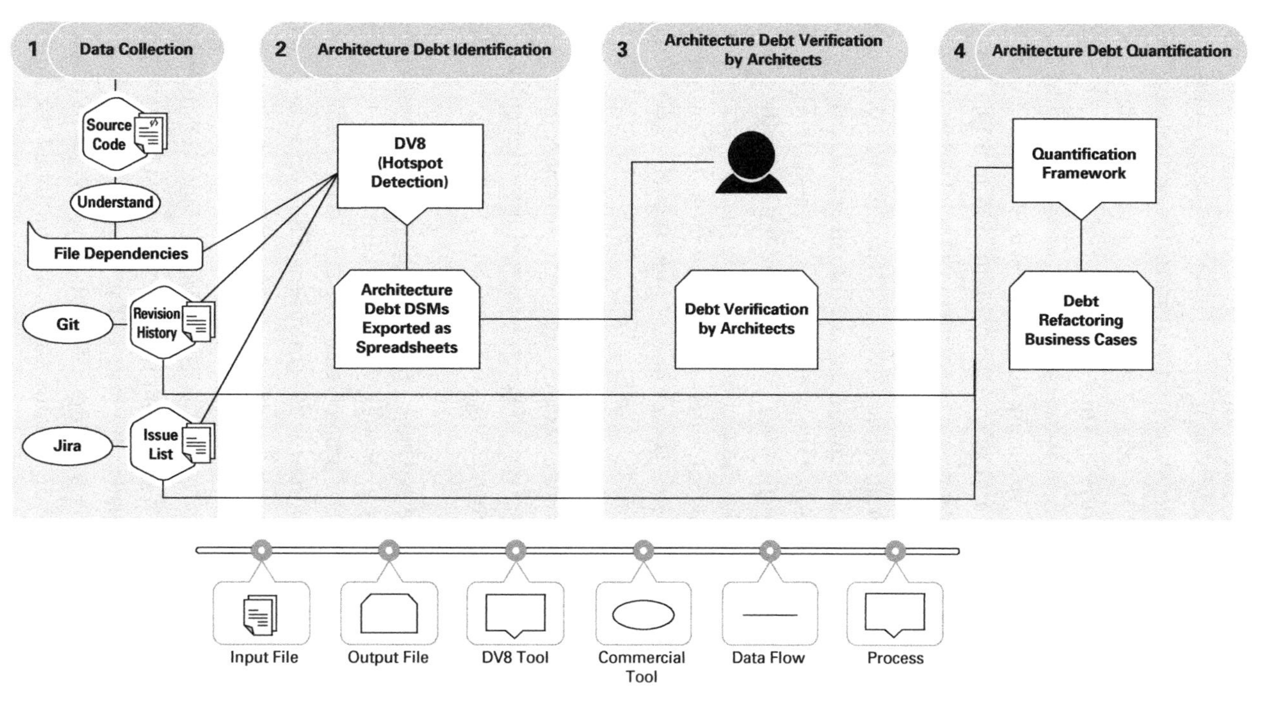

Figure 4.6
Data collection and analysis for the SS1 project.

To calculate the debt incurred by each of these DRSpaces we created a simple spreadsheet, as shown in figure 4.7. The three DRSpaces were identified by the names of the files that led them (anonymized to names of fruits and vegetables). As shown in cells A2–4, these three leading files were Pear.java, Apple.java, and Bean.java. They led groups of files of 139, 158, and 65 files respectively. However, some of the files were included in multiple DRSpaces. This is normal because a single file will often have many relationships with other files and so may be contained in multiple DRSpaces. To avoid double- (or triple-) counting the debt in such files we normalized the counts. Thus if a file appeared in two DRSpaces it counts as 0.5 in each. If a file appeared in all three DRSpaces it counted as 0.33 in each. In this way we arrived at normalized sizes of the three DRSpaces, as shown in cells C2–4. Then we counted the number of defects and changes incurred by each of these files over the prior year, and the lines of code committed to fix these bugs and make these changes; this is the *interest* on the debt. We could easily extract this information from the project's commit history and its issue-tracking system. These counts, appropriately normalized, are shown in columns E, G, and J.

Next we asked the architects to estimate the costs of refactoring these DRSpaces to remove the flaws. Because we could provide a list of each of the flaws and the files involved, the architects had little difficulty in making these estimates. The refactoring cost estimates, in person months, are given in cells K2–4.

Then we estimated expected bugs, changes, and churn for these DRSpaces after the refactoring. These estimates were based on average defect, change, and churn rates for the entire project (cells B11–13). The assumption here is that flawed files should revert back to being average files once the flaws were removed. Since, as we mentioned above, file bugginess, change-proneness, and churn have been shown to be highly correlated with architectural flaws, it is reasonable to assume that when the flaws are removed the files will revert back to being normal. This assumption was accepted by the SS1 project leaders and has been validated in practice. The expected numbers of bugs, changes, and churn for the files after refactoring are shown in columns L, M, and N. To arrive at these estimates we simply multiplied the number of files by the average bug, change, and churn rates in cells B11–13.

Finally we were able to arrive at an estimated ROI for the refactoring. The ROI is calculated as the expected savings (the profit of the refactoring

	A	B	C	D	E	F	G	I	J	K	L	M	N
1	***DRSpace Leading File***	**DRSpace Size**	**Norm Size**	**Current Defects/Yr**	**Norm Defects**	**Current Changes/Yr**	**Norm Changes/Yr**	**Tot LOC Changed**	**Norm LOC Changed**	**Refactor Cost (PM)**	**Norm Exp Defects/Yr**	**Norm Exp Changes/Yr**	**Norm Exp LOC Changed**
2	*Pear.java*	139	119.33	166	142.5	1068	839.2	49,171	42,213	5.5	39	346	20,281
3	*Apple.java*	158	133.83	63	53.4	607	451.7	25,603	21,686	7	44	388	22,745
4	*Bean.java*	65	37.83	72	41.9	429	207.2	17,807	10,364	1.5	12	110	6,429
5													
6	**DRSpace Total**		**290.99**		**237.8**		**1498**		**74,263**		**96.0**	**843.871**	**49,455**
7	**Project Total**	**797**		**265**		**2332**		**135,453**		**14**			
8	**Savings**										**142**	**654**	**24,808**
9													
10													
11	Base defect rates	0.33											
12	Base change rates	2.9										**Exp PM saved**	**41.35**
13	Base LOC/file	169.95											
14	Base /PM	600											

Figure 4.7
Debt calculation for the SS1 project.

investment) divided by the expected cost. The expected cost, as we mentioned above, was the cost of refactoring that was estimated by the architects. The expected benefit—the savings—is the number of lines of code that will not need to be written as a result of doing the refactoring. This expected value is calculated as the difference between the current churn and the expected churn. For the SS1 project the expected lines of code saved per year was 24,808, as shown in cell N8. This is simply the sum of the expected savings for the three refactored DRSpaces (cells N2–4).

Using the company-provided productivity rate (600 LOC per person-month; cell B14) a savings of 24,808 lines of code means a savings of 41.35 person-months per year. Given that the refactoring was estimated to cost fourteen person months (as shown in cell K7, which is simply the sum of cells K2–4), the estimated ROI of paying down this design debt, for the first year alone, was nearly 300%. This did not even take into account the reduced quality-assurance time and improved reputation that would come from dealing with far fewer bugs. This was a very easy decision for SS1's managers to make.

We provide another example of this type of analysis in Case Study A: Brightsquid and show how it paid huge dividends in terms of improving project productivity. This kind of automated architecture analysis is far more effective than simply calculating code metrics, which is what most companies do today (if they do anything at all) to manage code and design debt. (See box 4.4 for our advice on metrics.)

4.3 Avoiding Design Debt

The best way to avoid architecture debt is to never incur it in the first place. Many systems are created with no thought to architecture whatsoever. Indeed the Agile Manifesto claims, as one of its twelve principles, that "The best architectures, requirements, and designs emerge from self-organizing teams." Perhaps this works well for small systems, for systems where the design choices are few, or for noncritical systems, but for larger, less precedented systems we have never seen this work out. Many of the case studies in this book are a testament to the fact that design—good design—does not just emerge (see, for example, box 4.5).

Would you fly in an aircraft if you learned that its flight control system had simply emerged from a self-organizing team? We thought not. Emergence more commonly produces messes, like the ones depicted in figure 4.3. So if

Box 4.4
Software Metrics and Lines of Code

Software metrics have been around as long as the software industry. The predominant metric is lines of code, which has problems as a measurement construct: some developers write terse code, others take ten lines to write what could be done in two. Program size also depends heavily on language choice and framework use. However, at a coarse level, lines of code (abbreviated LOC) does give a rough sense of complexity—for example, by comparing orders-of-magnitude.

Other metrics have been proposed, including cyclomatic complexity (the number of paths through the program), coupling, and the ones we use in this book, like dependency counts, coupling measures, and hotspots.

Don't use metrics as a universal truth. They all have significant measurement error and precision issues. But do treat them as useful early warning indicators. Looking at trends, in particular, can rapidly highlight potential problems (e.g., a growing bug rate). The numbers in Figure 4.7 can give a false sense of precision. What is more important is the relative importance the numbers can help with: fix the problems that are more costly, and ignore the ones that seem trivial, regardless of the raw metric.

Box 4.5
Voice of the Practitioner: Marco Bartolini

The [SKA] project [a billion-dollar radio telescope], as of July 2019, is concluding [its] design phase. During this phase the telescope has been designed as an aggregate set of different elements, coordinated by a lightweight central organization. Among these elements, some are definitely software-heavy, such as the science data processor (SDP) and the telescope manager (TM). For these, we concentrated the design effort on the development and delivery of a well-documented software architecture, validating some assumptions via prototyping when possible in terms of time and resources. We are now in a phase where we are integrating these architectures into a single system of systems, incrementally developing an evolutionary prototype that contributes to validate and verify the architecture designed in the previous phase.

. . .

We always bear in mind that our software system is expected to be maintained and updated for the lifetime of the telescope, that is at least fifty years. This often ends in evaluating the tradeoff between the maturity of a product and its immediate performance benefits. Maintaining an open approach and adopting a set-based design is vital in making decisions at the most profitable moment.

—MB[6]

you are building a complex system, you might want to avoid at least some design debt by doing some up-front work on the design, perhaps employing a design method such as attribute-driven design (ADD). ADD emphasizes the use of *design concepts*—such as patterns, tactics, and frameworks—as a means of organizing your thoughts, limiting the space of design possibilities, and making competent design decisions.

Despite your best efforts and irrespective of how beautiful your architecture is at its inception, it is almost always the case that design integrity degrades over time, little by little, through the activities of adding features and fixing bugs. As we have said, these may accumulate into architecture debt. Once you have identified architectural debt, and once you have made a business case for its removal, you now need a strategy for removing such debt. This is typically achieved via refactoring.

The purpose of refactoring is to transform the structure of your system without (substantially) changing its functionality, to reduce coupling and increase cohesion among your system's components. This is, not surprisingly, one of the main goals of the move to microservices for many organizations. A system that is highly coupled is like a house of cards: it is hard to make any change in isolation and many changes will have negative ripple effects, causing the house to come crashing down.

To remove design debt, you frequently need to undo the effects of the antipatterns described above: unstable interface, modularity violation, unhealthy inheritance, clique, package cycle, and crossing. Fortunately, if these antipatterns have been identified this is not too difficult, because the antipattern itself is the guide to fixing the (self-imposed) problems. To fix a cyclic dependency or a clique you need to break the cycle, perhaps by moving code from one module to another or by reversing the direction of one of the dependencies. Fixing an instance of unhealthy inheritance is even clearer: move functionality from one or more child classes up to the parent. To fix a modularity violation, the implicit dependency needs to be made explicit. For example, perhaps there is an implicit assumption in a group of modules, about the length of a key or format of a file. If this assumption is made explicit and modularized, for instance put into its own module with an explicit (and abstract) interface that hides the details of the assumption, then the modularity violation magically goes away.

Each of these refactorings needs to be evaluated for its anticipated benefits and its costs, as we exemplified in the discussion of SS1 above. You need to remember that refactoring is not without its own risks. Bugs often go up, in the short term, immediately after a refactoring as you are (hopefully temporarily) destabilizing your system as you move it to what should be a more stable and more maintainable state in the long term.

4.3 Summary

In this chapter we have identified a number of root causes for, and kinds of, design (or architecture) debt. Unlike code-based technical debt, design debt is often harder to identify because its root causes are distributed among several files, and it is not detectable by static analysis alone. If you have a cyclic dependency where the cycle goes through six files, it is unlikely that anyone in your organization completely understands this cycle and it is not easily observable. So we often need help, in the form of tools, to identify design debt.

But this identification is worth it. Design debt can often undermine a project's very viability. We worked with one company, ABB, on a set of eight case studies, and they reported substantial design debt in six of those eight systems. One of the systems was so challenging—that is, it had accumulated so much technical debt—that the developers said that any change, no matter how seemingly trivial, was painful because no one understood the dependencies in the system. A change of a single line of code often took them a person-month of effort or more!

Once design debt has been identified, if it is bad enough, it should be removed through refactoring. In the ABB studies, all six of the identified systems are now undergoing major refactoring efforts, similar to what we detail in the Brightsquid case study (Case Study A).

Notes

1. http://lattix.com/lattix-architect.
2. https://www.silverthreadinc.com.
3. https://www.archdia.net/products-and-services.
4. https://bugs.chromium.org/p/chromium/issues/detail?id=58069.
5. https://archdia.com.
6. The full text of this and other Voice of the Practitioner sections can be found in the Appendix under the relevant name.

Further Reading

The definition of architecture (design) debt used in this chapter was borrowed from: L. Xiao, Y. Cai, R. Kazman, R. Mo, and Q. Feng, "Identifying and Quantifying Architectural Debts," *Proceedings of the International Conference on Software Engineering (ICSE) 2016*. The SS1 example was first described in: R. Kazman, Y. Cai, R. Mo, Q. Feng, L. Xiao, S. Haziyev, V. Fedak, and A. Shapochka, "A Case Study in Locating the Architectural Roots of Technical Debt," *Proceedings of the International Conference on Software Engineering (ICSE) 2015*. The notion of a design rule was analyzed and popularized by Carliss Baldwin and Kim Clark in their book *Design Rules: The Power of Modularity* (MIT Press, 2000).

A study that surveyed practitioners about technical debt found that architecture debt was the biggest source of debt in a project. This study is: N. Ernst, S. Bellomo, I. Ozkaya, R. Nord, and I. Gorton, "Measure It? Manage It? Ignore It? Software Practitioners and Technical Debt," *Proceedings of the 2015 10th Joint Meeting on Foundations of Software Engineering*, 2015.

Much has been written about the SOLID principles. The single responsibility principle, interface segregation principle and dependency inversion principle are attributed to Robert (Uncle Bob) Martin. You can read about it in *Agile Software Development, Principles, Patterns, and Practices* (Prentice Hall, 2003) on page 127. The open-closed principle is attributed to Bertrand Meyer in his book, *Object-Oriented Software Construction* (Prentice Hall, 1988), 23. The Liskov substitution principle was formed by Barbara Liskov (hence the name). It is described in detail in: B. Liskov and J. M. Wing, "A Behavioral Notion of Subtyping," *ACM Transactions on Programming Languages and Systems* 16, no. 6 (1994): 1811–1841. The GRASP patterns are described by Craig Larman in *Applying UML and Patterns* (Prentice Hall, 2004).

You can read about the historical roots of coupling and cohesion, along with many other principles of structured design in the seminal paper by W. Stevens, G. Myers, and L. Constantine, "Structured Design," *IBM Systems Journal* 13, no. 2 (June 1974): 115–139.

Some of the tools used to create and analyze DSMs are described in L. Xiao, Y. Cai, and R. Kazman, "Titan: A Toolset That Connects Software Architecture with Quality Analysis," *Proceedings of the 22nd ACM SIGSOFT International Symposium on the Foundations of Software Engineering (FSE 2014)*. The tools to detect architectural flaws are introduced in R. Mo, Y. Cai, R. Kazman, and L. Xiao, "Hotspot Patterns: The Formal Definition and Automatic Detection of Architecture Smells," *Proceedings of the 12th Working IEEE/IFIP Conference on Software Architecture (WICSA 2015)*. In addition, the consequences of design flaws have been analyzed in several papers, including Q. Feng, R. Kazman, Y. Cai, R. Mo, and L. Xiao, "An Architecture-Centric Approach to Security Analysis," *Proceedings of the 13th Working IEEE/IFIP Conference on Software Architecture (WICSA 2016)*; and R. Mo, W. Snipes, Y. Cai, S. Ramaswamy, R. Kazman, and M. Naedele, "Experiences Applying Automated Architecture Analysis Tool Suites,"

Proceedings of Automated Software Engineering (ASE) 2018. This latter paper includes the eight ABB case studies referred to in the conclusion of this chapter.

You can read about design methods, and the ADD method in particular, in H. Cervantes and R. Kazman, *Designing Software Architectures: A Practical Approach* (Addison-Wesley, 2016). ADD is a method for doing design in a repeatable way, employing design concepts such as patterns and tactics. If you want to read more about patterns and tactics you can find these in L. Bass, P. Clements, and R. Kazman, *Software Architecture in Practice*, 3rd ed. (Addison-Wesley, 2012) or in the multibook series on "Patterns of Software Architecture," beginning with F. Buschmann, R. Meunier, H. Rohnert, P. Sommerlad, and M. Stal, *Pattern-Oriented Software Architecture, Volume 1, A System of Patterns* (Wiley, 1996).

Finally, the original Agile Manifesto is at http://agilemanifesto.org/.

Case Study A: Brightsquid

Summary and Key Insights

In this case study, we look at a modern, product-focused software company and how it deals with technical debt. In particular, we show how the techniques from chapter 4, related to large-scale architecture analysis, can help a company with identifying and strategically managing its technical debt. We give detailed explanations on how to use design metrics to find problem spots in the codebase. In this particular case, refactoring the code to improve design problems resulted in more efficient implementation work. The metrics also helped the technical leads explain to management why this refactoring was important.

Background

Brightsquid is a software company based in Calgary, Canada, that makes secure communications solutions for the healthcare industry. The company has been in business for more than a decade and has been developing and evolving their core *platform* continuously over that time. Like most long-lived software, their platform has accumulated considerable technical debt. Brightsquid's developers felt this, but they had no way of identifying this debt, measuring its impact, and determining if it would be worth refactoring.

Analysis

We analyzed their platform using the DV8 tool suite, employing many of the analysis techniques described in chapter 4. Specifically we collected the following data for the Brightsquid platform:

- The dependency data extracted by the reverse engineering tool *Understand* from a single snapshot of the platform part of the project. This snapshot contained 1713 Java and Javascript source files.
- The Jira records over a period of eight months.
- The project's revision history, recorded as a Git log. These records covered the evolution history of the project over a ten-year period.

Based on this raw data we performed four different kinds of analyses:

- Calculation of architectural metrics
- Determination of architecture roots
- Calculation of architecture debt
- Determination of design flaws

We describe each of these analyses next.

Architecture Metrics

We first measured the overall level of coupling in the platform by employing two metrics: propagation cost (PC) and decoupling level (DL).

- *Propagation cost.* MacCormack's propagation cost metric—calculated based on a design structure matrix (DSM) representation of a system's dependencies—aims to measure how tightly coupled a system is on a scale from 0 to 1. Given a DSM of a project's dependencies, the PC metric process first calculates the transitive closure of the matrix to add indirect dependencies to the DSM until no more can be added. Given the final DSM with all direct and indirect dependencies, PC is calculated as the number of nonempty cells divided by the total number of cells. The PC metric therefore reports the average number of direct and indirect dependencies for all files in the project. This gauges, on average, how many files might be affected by a change to a single file. This is why the metric is called propagation cost—it indicates how likely it is for a change to propagate.
- *Decoupling level.* Decoupling level measures how well an architecture is decoupled into modules. DL is a score between 0 and 1. A system with a DL of 1 is totally decoupled (i.e., there are no dependencies between any two files). A system where a module influences all other modules (directly or indirectly) has a DL of 0. Of course, all real projects land somewhere in between. The point is that the more files that a given file influences, on

average, the lower its DL. In addition, the larger a module, the more likely it will influence more files, and hence the lower its DL. Conversely, the more that files are independent from each other, the higher the DL. Thus, the DL metric reports how *decoupled* a system's files are, on average.

Better architectures tend to have lower PC and higher DL, indicating better project-wide decoupling. For comparison, table A.1 shows the average levels of DL and PC for a benchmark set of 129 projects that we previously analyzed: 21 commercial and 108 open-source projects. (Interestingly, open-source and industrial projects do not differ substantially in terms of their PC and DL measures.)

The PC and DL measures for Brightsquid's platform are given in table A.2. As we just mentioned, for an architecture, a higher DL is better, and a lower PC is better. In this light, the Brightsquid platform's PC is low, and its structural DL is high, which seems quite encouraging. However, these analyses are based on purely structural information. We report two types of DL calculations:

Static dependencies: dependencies are extracted from code information (e.g., what module is imported, called, or used by another).

Static + historical dependencies: in addition to the above static dependency information, dependencies are extracted from version control history (commits). For example, if two modules are frequently modified together, we infer that there is a dependency between them.

Table A.1
Industrial benchmark data. Higher DL and lower PC are better.

	Open Source		Commercial		All Projects	
Stats	DL	PC	DL	PC	DL	PC
Average	**0.6**	**0.2**	**0.54**	**0.21**	**0.59**	**0.21**
Median	0.58	0.18	0.56	0.2	0.57	0.18
Max	0.92	0.72	0.93	0.5	0.93	0.72
Min	0.14	0.02	0.15	0.02	0.14	0.02

Table A.2
Architectural measures for the platform with and without co-change information.

	#Files	DL	PC
Platform (static dependencies only)	1,713	0.82	0.05
Platform (static+historical dependencies)	1,713	0.30	0.05

When history information is included in the analysis a different story emerges, as you can see from the second row of table A.2: the DL drops to a problematic 0.30. As you might recall from chapter 4, history (i.e., co-change) information is useful in design analysis because it can often reveal relationships between code files that are not apparent from any simple examination of the code or syntactic analysis.

For example, if two (or more) code modules share knowledge of a file format or a communication protocol, or any other implicit knowledge, then those files are related to each other despite having no overt dependencies. There are other kinds of hard-to-detect relationships as well: temporal dependencies (e.g., A must execute 50 ms. after B), semantic dependencies (e.g., A changes its behavior based on the mode of B), resource-based dependencies (e.g., A and B share memory or CPU or some other computational resource), and so forth. These dependencies are difficult to find but are often revealed in a project history as A and B, under these circumstances, they will often have been changed together to implement a feature or fix a bug.

Architecture Root Analysis

Architecture roots are the groups of files responsible for the most changes in the system. These architecture roots contain files (and their relations) that have the most impact on the maintainability of the system because error-proneness and change-proneness can propagate through their architectural relations.

Based on data extracted from the platform, we obtained the sets of architecture roots, that is, the groups of files that are responsible for the most maintenance effort in the project. In table A.3, we have listed the data of the four most important architectural roots detected. The meanings of each column of the tables in this section are as follows:

Table A.3
Architecture root analysis

	Root Size	Coverage	Cover_up_to
Root 1	128	0.23	0.43
Root 2	48	0.19	0.57
Root 3	50	0.50	0.68
Root 4	120	0.10	0.74

- Root size: the number of files within the root
- Coverage: the proportion of all bug-prone files in the project included in this root
- Cover_up_to: the proportion of bug-prone files covered by the top *n* roots cumulatively.

Table A.3 reveals the following facts about the platform:

- These four roots, cumulatively, cover 74% of the bugs in the project. The first two roots, containing just 128 and 48 files, cover 57% of the bugs in the project.
- Since the files within each root are architecturally connected, directly or indirectly, it is likely that their change-proneness propagates.

This result is consistent with all the other results that we have obtained from our analyses of hundreds of open-source and industrial projects: the majority of maintenance effort is always concentrated in a small subset of the system's files. In the next section we quantify the impact of these debts.

Architecture Debt Analysis

Architecture debt is defined as the extra maintenance effort caused by the flawed relations among *architectural roots*—groups of architecturally related files. In what follows, we quantify architectural debts as the additional numbers of changes and extra lines of code spent on the maintenance of the most flawed roots.

As we explained in chapter 4, the flawed architectural structures that propagate change-proneness or error-proneness to large numbers of a project's files are a kind of architecture debt. If these flaws are not removed, extra maintenance costs will accumulate in the form of *penalty*, or *interest*, on the debt.

In the platform, we consider each of the flawed file groups as an architectural debt and calculate its penalty, that is, the lines of code (LOC) changed and the number of commits made. These debts, if paid down, can result in significant return-on-investment (ROI). We summarize our data and analysis of the BrightSquid platform architecture debts in table A.4.

As can be seen in figure A.1, the top 4 groups, led by Contact_java, UserProxy_java, MessageAccount_java, and BrandName_java account for just 16.5% of the LOC of the project, but 79.3% of the bugs and 53.2% of the

A	C	D	E	F	G	H	I	J	K	L	M	N	O	P	Q	R	S	T	U
Root Index	Leading File	DRSpace Size	Norm Size	Current Bugs	Norm Bugs	Tot Loc bug changed	Norm Loc bug changed	Current Changes	Norm Changes	Tot Loc changed	Norm Loc changed	Norm Exp Bugs	Norm Exp Loc Bug changed	Norm Exp Changes	Norm Exp Loc Changed	Norm Extra Bugs	Norm Extra Loc bug changed	Norm Extra Changes	Norm Extra Loc Changed
root1	Contact_java	128	101	273	144	8,399	4,123	4,212	2,552	135,320	82,478	27	789	698	20,130	117	3,334	1,854	62,347
root2	UseProxy_java	48	47	55	53	2,000	1,993	1,127	1,069	37,069	36,002	12	363	321	9,245	41	1,630	748	26,757
root3	MessageAccount_java	50	31	241	114	7,275	3,391	2,773	1,210	79,879	31,625	8	238	211	6,081	106	3,153	999	25,545
root4	BrandName_java	120	105	134	48	3,718	1,223	2,714	1,447	82,280	38,492	28	817	722	20,826	21	406	724	17,665
	DRSpace Total		283		359		10,730		6,277		188,597	75	2,207	1,952	56,283	284	8,523	4,326	132,314
	Percentage		**16.5%**		**79.3%**		**80.3%**		**53.2%**		**55.4%**								
	Savings											284	8,523	4,326	132,314				
	Debt Percentage											**62.8%**	**63.8%**	**36.6%**	**38.8%**				

Figure A.1

Architecture debt calculation.

changes. If these roots were refactored, the bugs, changes, and churn in the project would drop dramatically. To refactor these roots, we propose removing their design flaws, as we discuss next.

Design Flaws and Their Severity Rankings

As described in chapter 4, the DV8 tool chain currently detects the following types of architecture flaws:

- *Clique:* a group of files that are interconnected, forming a strongly connected component but not belonging to a single module.
- *Package cycles:* typically the package structure of a software system should form a hierarchical structure. A cycle among packages is therefore considered to be harmful.
- *Improper inheritance:* we consider an inheritance hierarchy to be problematic if it falls into one of the following cases: (1) a parent class depends on one or more of its children; (2) the client of the class hierarchy uses/calls both a parent and one or more of its children, thus violating the Liskov substitution principle.
- *Modularity violation*: properly designed modules—ones designed with information hiding in mind—should be able to change independently from each other. If two structurally independent modules in a system are shown to change together frequently in the revision history, it means that they are not truly independent from each other. We observe that in many of these cases, these modules have harmful implicit dependencies that should be removed. We call this flaw modularity violation. In this project, since the number of changes and co-changes are few, due to the relatively short revision history provided, we consider two files to have modularity violations if they have changed together at least two times but have no structural dependency on each other.
- *Crossing:* if a file has many dependents and depends on many other files, then this file will appear to be at the center of a cross in its DSM. We call such a flaw a "*crossing.*"
- *Unstable interface:* if a highly influential file is changed frequently with other files that directly or indirectly depend on it, then we call it an "*unstable interface.*" In this project, we consider a file to be an unstable interface if it changes together with at least five other files two or more times.

Table A.4 shows the data of design flaws detected from Brightsquid's platform. The data shows a large number of *all* flaws, particularly modularity violations. But even ignoring those Modularity violations, there are large numbers of cliques, unstable interfaces, crossings, etc. These flaws have been shown, in studies of large-scale software projects, to be strongly correlated with increased bug rates, increased change rates, and increased churn.

Refactoring

Based on the design flaws identified on the platform, the Brightsquid team embarked upon a two-month refactoring. Refactoring to remove design flaws is actually quite straightforward. If there is unhealthy inheritance the refactoring typically moves one or more methods from the child to the parent class. If there is a clique, caused by a cycle of dependencies, then the cycle needs to be broken, usually achieved by removing or reversing one or more dependencies. If there is a modularity violation then the knowledge that is currently being shared across files needs to be modularized—that is, put into its own file, with an abstract interface, that the existing files can call. And so forth.

Of course normal development activities—implementing features and fixing bugs—did not cease during this time, and so the refactoring was done in parallel with the normal development stream. Once the refactoring was completed, another snapshot of the project was taken and analyzed.

In addition, to assess the consequences of the refactoring we captured an additional six months of project history. We now report on what we learned from analyzing the after refactoring version of the platform.

Table A.4
Design flaws in Brightsquid's platform

Architecture Flaw	#Files	#LOC Committed
Unstable interface	12	471
Unhealthy inheritance	60	222
Clique	17	71
Modularity violation	7,018	767
Crossing	29	387
Package cycle	34	242

Results

The results for Brightsquid were dramatic. They managed to reduce the number of files in the system, and the number of design flaws among those files, dramatically. Table A.5 shows the number of flaws before and after refactoring, as well as the numbers of files influenced by these flaws. For example, the number of Cliques went from 17 to 10, and the number of files involved in those Cliques went from 71 to 26. The number of Unstable Interface flaws went from 12 to 8 and the number of files influenced by these flaws went from 471 to just 59 after refactoring.

But, in the end, these numbers are meaningless unless they translate to productivity and quality improvements for Brightsquid. They did. Table A.6 shows how productivity changed after the refactoring, and the results were dramatic. In each case the totals of (nonbug) issues and bugs are for a similar six-month time period so that they can be fairly compared.

The amount of churn (committed lines of code) went from an average of 102 per commit before the refactoring to just 33.9 after refactoring—a reduction of 70%. And the average bug-fixing time went from 10.74 days per bug before to 7.31 days per bug—a reduction of 30%. In addition, before refactoring seventy-one change requests (including twenty-four bug reports), involving code changes in the platform were resolved in five months, whereas 150 change requests (including seventy-eight bug issues) were resolved in a similar five month period after the refactoring. Significantly, all of these improvements were achieved using the exact same team of developers.

The refactoring to remove architecture debt was seen as a major win for Brightsquid. Management felt that the combination of explicitly identifying

Table A.5
Architecture flaws in the platform before and after refactoring

Architecture Flaw	Count Before	Count After	Files Influenced by Flaw Before	Files Influenced After
Cliques	17	10	71	26
Unhealthy inheritance	60	30	222	102
Unstable interface	12	8	471	59
Crossings	26	6	387	47
Package cycles	34	19	242	94

Table A.6
Productivity measures for Brightsquid's platform

Measure	Before	After
Files	1,713	711
Issues opened	680	843
Issues fixed	583	653
Bugs opened	157	310
Bugs fixed	137	267
Bugs that changed code in platform files	24	78
Amount of churn per bug	102	33.9
Average bug fix time	10.74	7.31

debt items, in the form of architecture flaws, and quantifying the ongoing cost of that debt, was key to making the business case for refactoring. As the lead architect of the platform told the author in a personal communication from September 20, 2018: "Having an architecture debt analysis report that goes through coupling, circular relationships, and dependencies confirmed our hypotheses and we were able to convey to the top management that we need to do the refactoring as quick as possible."

Further Reading

Much of the experience and many of the results reported in this chapter were derived from: M. Nayebi, Y. Cai, R. Kazman, G. Ruhe, Q. Feng, C. Carlson, and F. Chew, "A Longitudinal Study of Identifying and Paying Down Architectural Debt," *Proceedings of the International Conference on Software Engineering (ICSE) 2019*. An earlier case study discussion how architecture debt can be detected and quantified can be found in: R. Kazman, Y. Cai, R. Mo, Q. Feng, L. Xiao, S. Haziyev, V. Fedak, and A. Shapochka, "A Case Study in Locating the Architectural Roots of Technical Debt," *Proceedings of the International Conference on Software Engineering (ICSE) 2015*.

The decoupling level metric was first introduced and empirically validated in: R. Mo, Y. Cai, R. Kazman, L. Xiao, and Q. Feng, "Decoupling Level: a New Metric for Architectural Maintenance Complexity," in *Proceedings of the 38th International Conference on Software Engineering, 2016*. A. MacCormack, J. Rusnak, and C. Baldwin introduced the propagation cost metric in "Exploring the Structure of Complex Software Designs: An Empirical Study of Open Source and Proprietary Code," *Management Science* 52, no. 7 (2006).

The most complete discussion of architectural flaws and its relationships to architecture debt can be found in: R. Mo, Y. Cai, R. Kazman, L. Xiao, and Q. Feng,

"Architecture Anti-Patterns: Automatically Detectable Violations of Design Principles," *IEEE Transactions on Software Engineering, 2019.*

Another interesting discussion of architecture debt was L. Xiao, Y. Cai, R. Kazman, R. Mo, Q. Feng, "Identifying and Quantifying Architectural Debts," *Proceedings of the International Conference on Software Engineering (ICSE) 2016.*

More information on the DV8 tool can be found at: https://archdia.com/.

5 Implementation Debt

> Code is like humor. When you have to explain it, it's bad.
>
> —Cory House

> It's important to have a sound idea, but the really important thing is the implementation.
>
> —Wilbur Ross

While design and requirements debt have far-reaching implications for product cost and quality, it is in the source code that many of these implications ultimately can be found. This chapter explains how to identify, manage, and avoid technical debt found in the implementation artifacts—source code—of the project. Two examples of how implementation debt comes to exist are prototypes that become a product and outdated technology choices.

Too often, what your developers wrote years ago as a prototype is still "the product" today. Because this prototype has not been updated and was written with expectations from another time, when security was merely a nice-to-have, it has hundreds of vulnerabilities waiting to be discovered by hackers. And because you used an old backend, it does not scale easily and your infrastructure costs are a financial burden for your company.

Because your project was written in an old language or framework nobody is using today (such as COBOL), nobody really understands it completely, not even the most senior developer on your team. And so no one wants to take on the risks of updating it and fixing existing issues. The consequences of keeping and maintaining such a code base are inevitable: code changes take longer than expected because it is difficult to understand the implications of a change, and it is hard to test and validate.

Writing good code is like staying healthy: the rules are simple—exercise and eat small portions of nutritious food—but enforcing these rules consistently for a long time is hard, and most people just don't do it. If you do not follow these rules, you will almost certainly accumulate some sort of debt. Once you do not follow the rules for a few months, it is usually painful to pay back the debt. Fail to follow them for a few years, and it will be almost impossible for you to pay it back, as we explained in chapter 1. Since we can seldom predict what parts of the code base will be important in the future, consistently and systematically writing good code from the beginning is important. Writing good, maintainable, and testable code should be your everyday coding diet. The not-so-important code you write today is tomorrow's business critical component.

Take the example of Twitter that used the Ruby on Rails framework in its early days (which we will discuss in Case Study B). As the company continued to grow, they did not change their underlying stack and hence they ran up against the limitations of this framework: the only way to scale was to deploy more CPU resources (hence incurring substantial additional costs) and to cache database results as much as possible (which is difficult to achieve in a consistent and timely manner). Needless to say, users felt the limitations of this approach quickly and the site was famous for being down in its early days. Since then, the company addressed these technical challenges and has paid back its massive technical debt. Note that this example showcases how requirements (the need to scale), design (the choice of Rails), and implementation all interact. While our chapters cover them as separate issues, the concerns are cross-cutting.

This chapter, covering those aspects of debt that relate to implementation concerns, follows the same pattern as the other chapters in this book:

1. **Identify** the constructs and patterns that constitute technical debt in code.
2. **Manage** your technical debt in code, by deploying appropriate tools and metrics that can help you manage technical debt in your code.
3. **Avoid** them down the line by instituting the processes and using the right tools to ensure that you don't have technical debt in your code in the first place.

5.1 Identifying Technical Debt in Your Code

Technical debt in code comes from taking shortcuts in how code is created at your organization. While some shortcuts are deliberate (e.g., duplicating code to get a new feature out), we discuss those that tend to occur frequently and inadvertently. Think of these as similar to a chef who fails to clean and put away their pots and knives. Implementation debt we discuss includes coding style, inefficient code, deprecated code, and code duplication.

5.1.1 Coding Style

There are two dimensions of coding style that you need to consider:

1. **Syntactic:** what the code should look like
2. **Semantic:** what features of the language to use or to limit

Syntactic aspects What the code should look like is not merely a cosmetic feature—it has important consequences in terms of readability. Making the code readable facilitates peer review and maintenance, which accounts for between 50% and 80% of software lifecycle costs. For example, the type of indentation (tabs vs. spaces), how many spaces per tab, where to insert braces, and so on. This sounds trivial to most managers but reading a codebase with mixed indentation and strange placement of braces can be a nightmare. Some languages (such as Python) require the developer to indent their code in a particular way, which is a good thing, and some come with automated formatters (like gofmt for go). But others do not, and poorly indented code can, for example, cause you to miss a condition. Good examples of how to approach (and enforce!) the cosmetic aspects of the code can be found in the NetBSD coding style or the Linux kernel coding style.

Semantic aspects How to use or restrict language features is as important as code style, perhaps even more so. Languages often provide you many different ways to express a concept, and using mixed approaches to do the same thing is confusing. Your code base needs to be consistent, and your team needs to use only one approach to perform an action so that it is the same through your code base. This makes code review and code understanding (for debugging and modification) much easier. For example, to define a constant in C, some will use a `const` variable while others will use a macro (`#define`). There are pros and cons for each solution (that are beyond the topic of this book), but really just one thing matters: establish one rule for your project and stick to it. It will make the codebase clean and consistent, which will improve the readability and understanding of the codebase, minimizing manual coding mistakes down the road, and hence minimizing (this aspect of) technical debt.

For some reasons (clarity, removal of ambiguous and complex features) and depending on the language, you also might want to avoid the use of certain features. For example, in C++, you might want to remove the use of raw pointers and only use smart pointers (that define a clear scope across the code base). For Java, you might allow (or disallow) the use of certain libraries or some functions. If you are developing a critical, real-time system you probably need to avoid any memory allocation at run-time. Language feature restrictions depend on your product constraints (e.g., web, real-time, embedded, portability concerns, etc.) and you must clearly define coding guidelines that state what is allowed or not.

What is important here is to enumerate, formalize, and *explain* why the codebase must comply with these rules. If you neglect to justify why—and this should be done in your coding standards—developers will argue with the rule and proceed in their idiosyncratic fashion.

5.1.2 Inefficient Code

Some code patterns are costly and inefficient, slowing down your program. This is very common in codebases that have been built over several years, and some tricks have been used to fix emergent issues, or perhaps when novice programmers implement a new feature without careful review. One common occurrence of such inefficient operation is string concatenation, and this can occur in almost any language.

Consider the following program in C++: it just concatenates the string "`bar`" to the string "`hello`" 1,000 times. This is a very simple program—essentially the same type of program we find today in many codebases to build a URI or generate output (such as JSON or XML documents).

```
#include <string.h>
#include <iostream>

using namespace std;

int main() {
   string foo("hello");
   for (int i=0 ; i<1000 ; i++)
```

```
        {
            foo=foo+"bar";
        }
        cout << foo;
    }
```

The program does the job but let's try to evaluate the performance of such a program[1] using Valgrind[2] (a memory leak detector and profiler).

```
==17081== HEAP SUMMARY:
==17081== in use at exit: 0 bytes in 0 blocks
==17081== total heap usage: 1,995 allocs, 1,995 frees,
4,586,137 bytes allocated
```

We find that for a simple block of code like this, you perform 1,995 memory allocations. How is that possible?

The answer resides in the following line:

```
foo=foo+"bar";
```

When we do this, the old foo variable is being destructed (deallocated) and a new one is allocated with the content of the previous one concatenated with the string "bar". For a simple concatenation, the program pays a huge price in terms of memory allocation. This is highly inefficient and if this function is being called frequently, it will have a serious negative impact on performance. Of course, when this function runs on your laptop, you do not see the cost of such inefficiencies, but when you are running multithreaded programs handling thousands of concurrent requests with hundreds of functions like this, the performance cost is real. This constitutes technical debt in the sense that scaling the application will require more computing power, which will ultimately have technical consequences (scaling is known to be hard) and financial consequences (the cost of acquiring more processing power, which is the interest on this particular debt).

This type of inefficiency is common. Fortunately, many languages provide efficient solutions. For the string issue, the C++ standard library provides the stringstream type, which preallocates memory to store a string and just concatenates until it reaches capacity (and reallocates memory when there is no memory left). The previous program can be reimplemented with stringstream and will look like this:

```
#include <string.h>
#include <iostream>
#include <sstream>

using namespace std;

int main() {
  stringstream ss;
  ss << "hello";
  for (int i=0 ; i<1000 ; i++)
  {
    ss << "bar";
  }
  cout << (ss.str());
}
```

Let's run Valgrind again; we notice that this is much better—in fact, two orders of magnitude better—in terms of memory allocation as it does far fewer memory allocations and frees.

```
==17098== HEAP SUMMARY:
==17098==   in use at exit: 0 bytes in 0 blocks
==17098== total heap usage: 7 allocs, 7 frees, 84,418 bytes
allocated
```

This type of inefficiency is found in almost all languages. In Java, developers will concatenate `String` variables while the alternative is to use `StringBuilder` or `StringBuffer`. Replacing a string concatenation operation by a sequence with `StringBuilder` and `StringBuffer` can lead to a significant performance improvement. (Note: If you are wondering about the difference between these two types, `StringBuilder` is not thread-safe, whereas `StringBuffer` is). Also, the compiler can optimize your program and operations such as these can be optimized behind the scenes. Depending on your language and compiler settings, this inefficient code may be optimized and run as fast as the most efficient versions. However, it is better not to rely on your compiler for this, but instead address these issues in the source code itself. Static code analysis tools can easily detect this common pattern (more on this in the Managing Implementation Debt section).

5.1.3 Use of Old, Deprecated, or Insecure Functions or Frameworks

One of the most notable examples of technical debt is any C code that still uses `str*()` functions. These functions have been known for years to have security impacts and are famous for being the original root cause of buffer overflows. Basic secure coding principles recommend simply banning the use of such functions.

Similar issues are also present in the web-development world, where developers build their own SQL requests based on user arguments that are not filtered or that are filtered in ad hoc ways. This well-known SQL injection issue is responsible for countless numbers of attacks and data breaches. And solutions have existed for years against these kinds of bugs, such as using an intercepting validator pattern.

Using such outdated functions is an example of acquiring technical debt: by not using more modern and secure alternatives, you avoid the near-term costs of making changes to your code and avoid the costs of refactoring, but you expose your product to potentially disastrous security defects. The examples with `str*` functions in C are well-known, but each language has similar issues. If your codebase uses these old, insecure functions, just add an item in your issue tracking system to remove them. And, ideally, replace your custom security code (e.g., for input validation on a web form) with the services offered by a security framework (discussed further in box 5.1). Every respectable web-development framework has its own validation tools.

Box 5.1
To Use or Not to Use a Security Framework

A few years ago I was involved in a study of the approaches to security in four web-based systems. In each of these systems we attempted to assess the consequences of their approaches to security. Some of these systems implemented their security functions directly in their code base. Some adopted frameworks. Some did a blend. As you might guess, in the systems where they attempted to write their own security controls, the results were a disaster. These systems had many high-risk vulnerabilities (as determined by an automated scanner), and they were spending about 20% of their total project effort on security. On the other hand the systems that adopted (or refactored their systems to adopt) a frameworks-based approach to security did far better on the scans and spent less than 10% of their project effort on security.

Refactoring (to reduce technical debt, or for any other reason) should always be justified in terms of return on investment. Here the return on investment was stunningly clear. It pays, in terms of both effort and risk, to refactor your code to adopt security language features or framework features.

—RK

5.1.4 Code Duplication: Just Stay DRY

One common coding issue that incurs technical debt surrounds code duplication. Copying and pasting existing code provides the illusion of a quick fix, but it will eventually produce spaghetti code, hard-to-understand code, hard-to-maintain code, and many other forms of potential debt. There is, however, a programming best practice that applies here: don't repeat yourself (DRY). By staying DRY, you do not replicate code, and hence you do not replicate potential errors. One buggy, nonrepeated line of code used by ten functions is fixed by a one-line edit. The same line that has been copied ten times will require ten edits and will waste much more than ten times the developer's time to figure out where it has been copied, and then to make the change uniformly. If you do not follow the DRY principles, you will waste everyone's time (WET). In the previous chapter this phenomenon was identified, at the design level, as an architectural flaw called *modularity violation*.

Code duplication almost always follows the same story: a developer tries to implement something, and the solution to their problem already exists in some part of the codebase. The developer assumes that, as the existing code already works, the best thing to do is just to copy the code, tweak it a bit, and reuse it. Simple! As development proceeds, different programmers do the same thing in their own contexts, and you end up with the same piece of code (partially) replicated in dozens of places in your code base. On a macro level, the same thing happens with copy-and-paste of code blocks from StackOverflow or other social coding sites. But one day, somebody discovers a bug or a security issue, or just wants to make a change to the logic in the original piece of code. In this case, the complete fix will require changes to the original code and all of its replicas. As time passes, the original developers that duplicated the code have moved on (or merely forgotten what they did four years ago), and the developers currently working on the product are not sure how to address the change, which delays the release of the change, decreases your confidence that you changed all of the places that needed to be changed, and likely increases the number of commits you are going to have to make. Not good!

Let's consider a concrete example: OpenSSL CVE-2006–2940. This bug was a denial of service (DoS) vulnerability that was caused by an attacker employing public keys with very large exponent and modulus values in X.509 certificates. These large values required an enormous amount of processing, which led to the DoS attack. The simple fix to OpenSSL was to set a maximum size for these values and check the packet size before processing it. This "simple" change, however, affected four different crypto algorithms within OpenSSL: DSA, RSA, DH, and EC. The developers had the same, or similar, code in each of the algorithm implementations (duplicated code), and these were implemented in three files per algorithm. The developers therefore needed to examine all twelve files and find and fix all of the affected code (see box 5.2 for more on duplication).

The remedy to such a problem should follow this story: A developer detects the problems and finds an appropriate solution in their code base. They refactor their code, put the common code into a single module that is shared across the project, run tests to verify that no regression bug has been introduced by this change, and write new code that will use this new module. When a bug/vulnerability is discovered, only the common module needs to be fixed. Releasing such a fix will typically take a few hours or a few days at most. This is how software engineering should be done. And this is where coding meets architecture. Rather than taking a coding approach to fixing a problem—finding the affected code in each of its incarnations and fixing it then and there—we advocate taking an architectural approach, factoring out the problematic code and modularizing it.

Box 5.2
Code and Bug Duplication

A long time ago, I was working on a modeling platform written in Java. We developed code that was used not only by research institutes but also by customers in industry. One of the developers was copying and pasting code for almost every new feature. That created chaos over time, and almost any change in the codebase broke not only the new functions but also the older ones.

Over a few years, the number of duplicates increased by 30%! As we had few tests, many bugs were discovered postrelease, which forced us to release emergency fixes and other patches. Of course, this reduced our productivity (when we were engaged with these fire drills we were not working on new features) and also increased customer frustration ("Why the hell is my software no longer working!") and decreased confidence ("How can I trust this software to produce the right results?"). These issues could have been fixed by avoiding code duplication in the first place and by writing properly modularized code.

—JD

5.2 Managing Implementation Debt

The previous sections defined some coding patterns or processes that may incur technical debt. In this section, we will explain how you can start to manage them.

5.2.1 Use Code Analyzers

Static analysis to the rescue Static analysis tools process code and discover syntactic and semantic problems. They detect a wide range of issues, from simple indentation style (consistent use of tabs, spaces, etc.) to potential security issues. For some languages, they also surface the use of deprecated features or methods. These tools will produce a list of violations that might vary in terms of severity, and you need to start addressing the most important first (e.g., anything related to security or safety) and then, as time permits, start to look at the minor ones (such as code formatting).

There are now static analysis code tools available for all of the most popular programming languages. As these analyzers look for different defects in your codebase, it is important to not rely only on a single tool but rather to use multiple tools to increase the list of defects that can be detected. Since some tools focus on security issues while others will focus on code formatting, increasing the number of tools will give you more insights about your codebase and, ultimately, help you prioritize what needs to be fixed. Also, to prevent information overload, customize your tools to filter out trivial results.

Check for duplicated code Looking for code duplication is not as straightforward as it sounds. A very naive (albeit a good point to start—better than nothing!) is to only look at individual lines that are similar. Such an analyzer acts at the level of raw text and does not take into account language semantics.

We usually distinguish several types of duplicated code:

1. Syntactically similar with only minor variations in terms of whitespace and layout
2. Syntactically similar with just changes to identifiers and types
3. Syntactically similar with some updated or inserted statements
4. Semantically similar but syntactically different

Level 1 of duplication is straightforward to detect. Alas, when programmers duplicate code, they often make small changes (e.g., change a variable, class, or function name) and a naive code duplication detection tool fails to detect such issues. The following table shows a very simple (and inefficient) code example that will fail to be detected by a naive code duplication tool.

def checkValue(value:
String) {
if (value == "foo"){
return True;
}
return False;
}
	def checkValue(value:
	String) {
	if (value.
	equalsIgnoreCa
	se("foo"))
	{
	return True;
	}
	}

Original code	Duplicated code

More effective code duplication tools use more advanced techniques, either by employing string processing algorithms (such as Rabin-Karp) or through the use of language semantics to look for similar code patterns in the Abstract Syntax Tree (AST) of the language. These problems are very similar to detecting plagiarism in multiple textual documents. Such tools are far more difficult to implement but provide more accurate results.

As for static analysis, there are several open-source and commercial tools available. They mostly rely on advanced string processing and implement the Rabin-Karp algorithms. These tools tend to provide accurate results. They also need to be configured. For example, you can specify a minimum number of tokens or lines before considering a block of code a duplicate, maximum numbers of duplicates, and so on. When beginning your hunt for duplicates, start with a high value for the number of lines/tokens (e.g. 100 lines of code) so that you can focus on large blocks of code that have been duplicated. Once these blocks have been refactored, decrease these values to focus on other, smaller duplicates (e.g., twenty lines of code).

Vulnerability scanner Security is so important that there is now a vibrant market of static analysis tools dedicated to detecting security issues in code. Such tools are often specifically looking for documented security problems—such as Common Vulnerabilities and Exposures (CVE) or Common Weakness Enumeration (CWE)—which are security issues that should not be in your codebase. The techniques underlying these tools overlap with static analysis tools as they are attempting to detect code patterns related to CVEs and CWEs. While most of the tools initially focused on unsafe C functions, there are now tools that also detect issues in web-based applications (and detect potential vulnerabilities such as cross-site scripting or SQL injection).

Such vulnerability scanners should be used in your project, especially if you develop a product that is exposed to the public (either an embedded system or a web service exposing a public API). Some static code analyzers also check for CVE and CWE violations. In the open-source world, flawfinder is a very good tool for C/C++ code that can surface many security issues.[3] Left untreated, such vulnerabilities can expose customer data or cause your service to go down for an extended period of time. Repaying this debt early and proactively helps to avoid these issues.

5.3.1 Define Your Coding Guidelines

To keep the code consistent across your project and to make sure that syntactic and semantic rules are enforced, you need to define coding guidelines.

The coding convention should be located in a file with an explicit name (such as `CODING_GUIDELINES`) at the root of your project hierarchy and should be regarded as the single source of truth for code formatting/style/language restrictions. Reference this file in the guidelines that you give to any new developer when they join the team. Do not copy this information anywhere, just reference it (a copy will just duplicate the rules and create the same problem as duplicated code if and when you want to change these rules). Keeping a single copy to reference is part of good documentation practice, which we cover in chapter 8.

As for enforcing the policy, you have several different options:

- Some compilers are either very strict on the code structure or have options to enforce rules on the source code so that any noncompliant code will not compile. For example, GNAT for Ada has many language features and enforces the GNAT style at compile-time.[4] GCC/clang enforce some restrictions as well.
- Other languages (such as Python), have good tooling (e.g., flake8[5]) to check compliance with a coding style such as PEP8 (the coding style for Python).[6]
- Commercial static checkers can be configured to enforce your coding guidelines.

- Write your own static analyzer—probably costly in terms of time and maintenance. You will likely write something that is not as complete or powerful as existing tools. However, it may be more contextually relevant to your domain and problems.

It is important to strictly enforce the coding style and refuse any code change that does not comply with it. This validation needs to happen at code review (see the Avoiding Implementation Debt section of this chapter) or when a change is pushed. This might be frustrating for your developers at first, but it will pay off handsomely over the long term, in terms of consistency, understandability, and lower maintenance costs.

5.3.2 Use New Language Features

Languages and frameworks change and improve over time, and not taking advantage of these improvements can impact readability, security, performance, or all of the above.

Let's take the examples of C++ and Java. A few years ago, these languages (C++ pre-C++11 and Java pre-version 5) did not support the ranged-based for loop, which made the code more complicated to read. The code snippets below show the difference between Java pre- and post-Java 5. The code on the left shows iteration over a collection without a ranged-based loop (pre-Java 5), while the code on the right shows iteration using ranged-based loop (starting in Java 5). The right side is not only more readable but also less error-prone (e.g., one might inadvertently insert the line `list.get(j)` in the code on the left side, with `j` being defined somewhere else, and this erroneous code will compile and execute).

```
for (int i=0 ; i<list.
size() ; i++) {
System.out.println(list
.get(i));
}
```

```
for (String s: list) {
System.out.println(s);
}
```

As programmers use languages and as new revisions are being released, there are improvements that makes your code more readable, safer, and more secure. The language that has likely changed the most over the last decade is C++, with large numbers of new features: C++14 is almost a totally different language from C++98. Among the best new features are smart pointers (that facilitate memory management), lambda functions (that allow you to write more concise code), and automatic type inference. Similarly, Java has seen many improvements in the past decade—introduction of ranged-based loops, and the use of stream or lambda functions. Using these new features makes your code more robust and avoid potential technical debt. For this reason, you should try to use these features shortly after they are released and officially supported by your build system.

Similarly, you should pay attention to the frameworks (e.g., database or web frameworks) and external libraries you are using. Not only do new versions typically provide more functionality, but they also fix security issues (e.g., any programs that did not update their OpenSSL libraries after the discovery of the Heartbleed bug were exposed to serious vulnerabilities). Developers and managers should stay informed (by subscribing to mailing lists and visiting the websites of the products they use) about new or upcoming versions and start using them as they are released as stable. The objective is not merely to use the new attractive features just released (e.g., do not use development or testing branches in production) but rather to stay current, ensuring that the current codebase runs on a recent stable release that is up to date with performance and security fixes. These decisions should also be carefully made, as new versions might also not be backward compatible and require work before being adopted and deployed in production.

5.4 Avoiding Implementation Debt

Earlier in this chapter we explained not only what constitutes technical debt in your code (e.g., coding style, duplicate code, etc.) but also how to identify them (code analyzers) and what can help you to stay consistent and write good code. The next section explains what processes and tools can be used not only to avoid implementation debt but also make sure no more debt is being added in your code base.

5.4.1 Choose Your Language and Libraries Wisely

The language and libraries you choose will influence how easily you can manage your technical debt and, when starting a new project, they should be considered wisely. The selection of the language should be done carefully according to your domain and industry. Such decisions can have long-term impact, as described in box 5.3.

Similarly, library selection matters a lot, and even more today in specific domains. Often, developers are tempted to select the newest libraries, believing these will either help them to develop faster or integrate new features. Unfortunately, this is short-term thinking and we often overlook the possibility of the library being discontinued or no longer supported. In such a case, you will either have to migrate and use a new library or support the library yourself.

As for choosing a language or library, the best recommendation would be to monitor the usage in your industry. Signals such as the number of uses on GitHub (number of forks or likes), and the number of questions on StackOverflow will help you gauge the popularity for a given language or tool.

For example, if you are in charge of the design of a new website, C might not be the best language to use (and there are very few libraries to do so). On the other hand, Java for the backend and JavaScript for the frontend seem like great language choices. Both are used for this type of system and both have well-supported libraries (e.g., Play for Java, and React for JavaScript).

Box 5.3
Voice of the Practitioner: Nicolas Devillard

Reimplementing an existing piece of software is a lot more work than it seems. As a software engineer faced with a wall of legacy tech, you may be thinking, "Gee, these are thirty-six classes dedicated to doing something that is natively provided in Go. There are countless bugs in there, and I am not even sure I understand all the details. As soon as we move to Go this is all going away."

This is true, but you forgot that the legacy tech is in use by 200 customers around the globe. Come over with your shiny new Go version and they will ask you, "What's in it for me?" If your new version offers the same features, your customers have no reason to upgrade, and you will end up having two versions to maintain—the legacy one you wanted to get rid of, and the new ones with its new bugs.

In a previous company we had planned to replace four legacy products with just one, but ended up with 4+1 products: the legacy versions and the new product. Debt is sticky, it can stay for extended periods of time before it can be replaced with shiny new tools. There is a reason why the banking world is still relying on heaps of COBOL software.

—ND[7]

5.4.2 Efficient Code Reviews

If you are not familiar with code reviews, you should start adopting them today. Code review is one of the best ways to ensure that technical debt is not inadvertently accumulating in your code. While tools automate the detection of some debt, they cannot catch everything and ultimately it is crucial to conduct a review by human developers. But, to be effective, code reviews need to follow a well-defined process to make sure that all the rules are checked and the code is compliant with your coding guidelines. We present a set of rules to make code reviews effective and productive.

Rule 0: Automate basic checks Many aspects of a code review can be automated by using static analyzers that will check a code change against a set of rules. Depending on how the tool is integrated with your code review process, it can automatically add comments and also allow/block a new change. These checks should be done according to the coding guidelines (as discussed in the previous section of this chapter).

Automating such aspects of the code review allow developers to focus on other validation that cannot be automated. These less automatable aspects include validating the overall architecture, ensuring correctness of the code, test coverage, and quality of comments. This also increases developer productivity because they do not waste their time reviewing code that is not (yet) ready for review. Some code hosting platforms (such as GitHub) offer tools that automate the code review process.

Rule 1: Two reviewers per change One of the main purposes of a code review is to have a pair of fresh eyes on the code to look for traps and pitfalls that the author may have fallen into. Each developer will have their own biases, blind spots, and areas of expertise. Thus, using two code reviewers maximizes the chances of catching potential mistakes. Having two reviewers also avoids the *buddy ship-it* syndrome, where a developer finds somebody who is willing to ship *anything* and gets everything shipped regardless of quality.

If the code has been written during a pair programming session, the developers that participated in writing the code cannot review or ship the code. As they participated in the writing process, they are already too biased to have an honest, independent opinion about the code's quality.

Rule 2: Limit the scope of a code change It is often tempting to extend the scope of a code change: while developing a feature, we can also fix previous issues or rearchitect some components. The implementer has good intentions here (to improve the code base by making it simpler and safer), but this is generally a bad habit.

For one, it will include more changes in the review, growing its size and making it more difficult to review. A change with a well-defined scope and few lines should take ten to twenty minutes, at most, to review. A review that introduces an unscoped refactoring or rearchitecting might initiate discussions, taking more time to evaluate. This decreases development agility (the change will take longer to ship) and, while being discussed and reviewed, this change might need some updates (rebasing on top of changes that have been shipped in the meantime).

In addition, adding multiple code changes in the same review makes it harder to roll back if the change introduces a new issue. If a change is small and restricted to one single problem, it is easy to roll back and remove from the code base. If the same commit contains multiple changes at once, removing this commit will also mean removing all the changes that have been bundled.

For these reasons, code changes should be well scoped and small—as small as possible to achieve your immediate need. A code review should be a few lines of code (most of the time, under 100) so that it can be reviewed by peers in a few minutes. Large code changes should be divided into smaller code units, each one being reviewable in a few minutes by peers. In that way, changes are being shipped quickly and integrated into your master branch as fast as possible.

Rule 3: Define the commit message format The commit message should be as informative as possible so that reviewers know what to expect and question. The message should contain:

- One line that concisely describes the change
- A larger description that gives more details about what has been changed, why this change was introduced, and the approach that has been taken
- Reference to the issue on the issue tracker (a link or a unique identifier following a consistent pattern for traceability)
- How this change has been tested (adding more unit tests, deploying the software in a testing/development environment, etc.)
- What documentation has been added (and add a draft result of the documentation)

This will give all the information required for the reviewer to evaluate the code and see if anything is missing (e.g., the reviewer does not need to ask how the change has been tested) so that reviewers can focus on the real issues.

An example of a good commit message is shown in Figure 5.1. It explains the reasons for the change, includes the name of the author, the reviewers, a reference to the bug that was fixed, and any potential related issues on the issue tracker. Test information is in the subsequent discussion.

Filling in this information should take no more than one minute when sending your code for review and saves a significant amount of time when a potential problem arises in production and recent changes are being investigated (especially when linking the review to the issue tracker). You should require your team to adopt a message commit format and enforce it in your continuous integration pipeline (e.g., a code change without a good commit message or lacking a link back to the issue tracker is immediately desk-rejected). More discussion of this issue can be found in chapter 3, where we discussed requirements traceability and technical debt.

Figure 5.2 is an example of a bad commit message: very brief description, and no connection to the issue tracker. Why has this commit been made? We do not know.

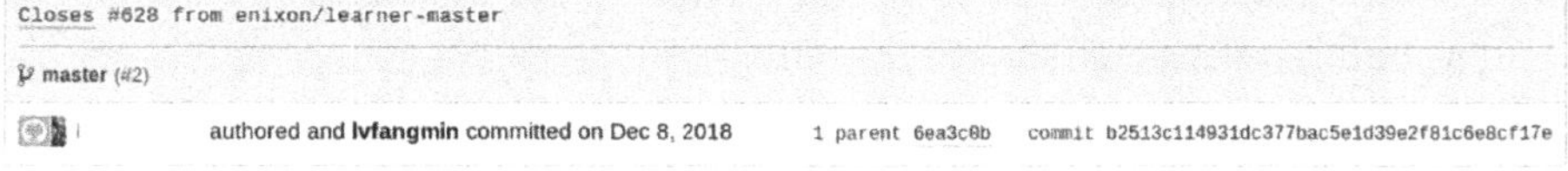

ZOOKEEPER-3140: Allow Followers to host Observers Browse files

Creates a new abstraction, LearnerMaster, to represent the portions of the Leader logic that are used in LearnerHandler. Leader implements LearnerMaster and a new class ObserverMaster implements LearnerMaster. Followers have the option of instantiating a ObserverMaster thread when they assume their role and so support Learner traffic.

A new parameter 'observerMasterPort' is used to control which Follower instances host Observers.

Author: Brian Nixon <nixon@fb.com>

Reviewers: fangmin@apache.org, hanm@apache.org, eolivelli@gmail.com

Closes #628 from enixon/learner-master

master (#2)

authored and **lvfangmin** committed on Dec 8, 2018 1 parent 6ea3c0b commit b2513c114931dc377bac5e1d39e2f81c6e8cf17e

Figure 5.1
Good commit messages explain the change and include other details.

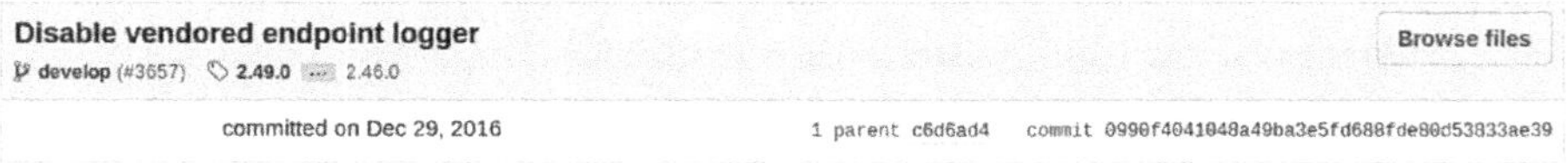

Disable vendored endpoint logger Browse files

develop (#3657) 2.49.0 ... 2.46.0

committed on Dec 29, 2016 1 parent c6d6ad4 commit 0990f4041048a49ba3e5fd688fde80d53833ae39

Figure 5.2
Bad commit messages don't explain the change.

Rule 4: Automate build and testing prior to ship Code reviews should be as easy and automated as possible. One common mistake that happens when there is a code change is to not check if the change breaks the build or fails to pass the tests in a clean environment. These activities should be automated by your code review system so that the author or the reviewers need not worry if the change breaks the build or introduces a regression. We go into more detail on the production environment in chapter 7.

In particular, your code review system should attempt to build the system and run the tests when the code is sent for review. If anything fails, the code is not ready for review and reviewers do not waste their time on it. Then, once the code is accepted, the continuous integration pipeline merges the changes, rebuilds the system, and reruns the tests to make sure that the changes have been correctly merged. This keeps the master/production branch consistent and operational. Make sure that you build and run tests against your production environment (that you correctly defined, as will be discussed in chapter 7). If you have several production environments or target platforms (e.g., if you develop embedded or mobile software) make sure you build and test against each of them.

5.4.3 Gather and Analyze Metrics on the Code Base

As you start using static analyzers to obtain a list of violations and duplicates, you need to analyze your trends. This analysis will show you if you have reduced the number of violations and if your refactoring has succeeded in removing duplicates and simplifying your code. To do that you need to collect code metrics and analyze them over time. Some interesting metrics to look at are changes in the number of violations per 1,000 lines of code, number of total duplicated lines, and number of total security or safety issues.

This can be done by building dashboards to visualize trends on your code base. There are online tools (such as Code Inspector or Code Climate[8]) that automatically analyze your code and show trends in numbers of code violations/duplicates. The following picture (figure 5.3) shows the dashboard provided by Code Inspector with trends, numbers of violations and duplicates.

The exact thresholds you want to meet depends on the industry you work in, the language you use and the type of system you are working on. For example, a safety-critical systems should target 0 defects while other systems might be more flexible.

(a) Key Metrics and Trends

	today	one day ago	three days ago	three weeks ago	three months ago
#SLOCS	293892	293892	293887	293976	N/A
Violations / LOC	0.0056	0.0056	0.0056	0.0053	N/A
Critical violations / LOC	0.0002	0.0002	0.0002	0.0002	N/A
Major violations / LOC	0.0008	0.0008	0.0008	0.0005	N/A
Medium violations / LOC	0.0010	0.0010	0.0010	0.0006	N/A
Low violations / LOC	0.0036	0.0036	0.0036	0.0040	N/A
Code Style violations / LOC	0.0033	0.0033	0.0033	0.0034	N/A
Error Prone violations / LOC	0.0000	0.0000	0.0000	0.0000	N/A
Documentation violations / LOC	0	0	0	0	N/A
Security violations / LOC	0	0	0	0	N/A
Design violations / LOC	0	0	0	0	N/A
Safety violations / LOC	0.0009	0.0009	0.0009	0.0002	N/A

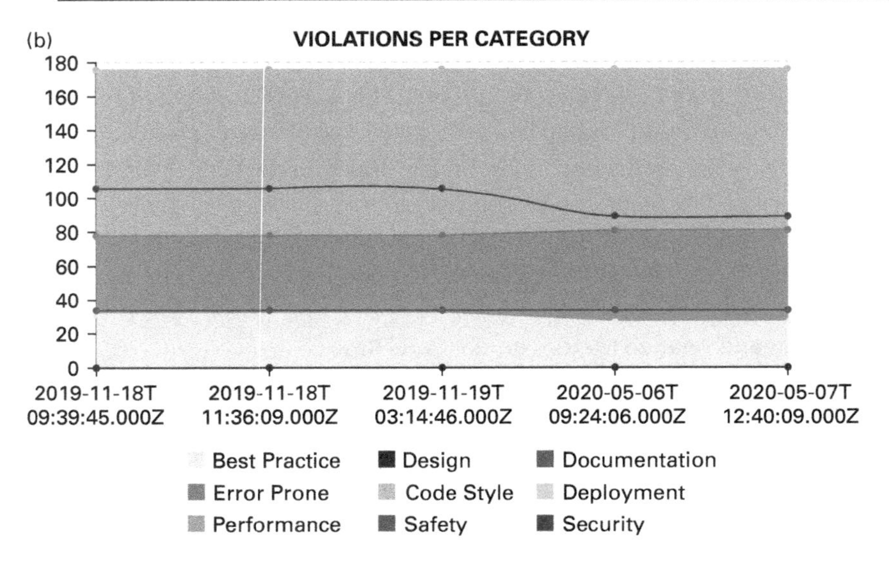

Figure 5.3
Dashboard from code-inspector showing trends in a codebase. (a) Metrics and trends; (b) categorized violations; (c) duplicated code.

By using such tools, you can verify that your code base is improving over time and that you are reducing the number of violations or duplicates. Removing technical debt on large projects takes a significant amount of time, and therefore it is important to keep historical data and track progress over time. We provide a case study showing an example of how this can be done in practice, along with the benefits that accrue to it, in Case Study A (BrightSquid).

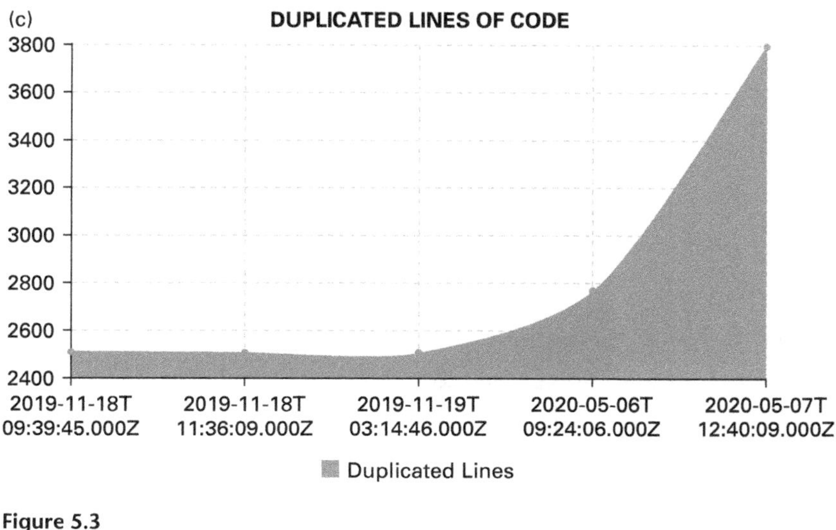

Figure 5.3
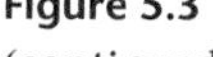
(continued)

5.5 Summary

Much of your technical debt takes root in your code. Language choices, design pattern use, inefficient language feature use, code duplication: there is an endless list of code issues that can cause you to incur technical debt. In this chapter, we have presented the most common causes of technical debt in your implementation and explained how to address them.

The most important guideline is to write down coding rules and make sure your team sticks to them. That means having the necessary tools that check the compliance of the code against these rules every time a line of code is changed (using a continuous integration pipeline that runs the checks on your code, as we will discuss in chapter 7).

One thing to keep in mind is to stick to the KISS (keep it simple, stupid) principle. Today, many languages have complicated and advanced features that might sound appealing for experts (e.g., C++ or Scala, to name just two languages, have many complex features). But this complexity may be a blocker for newcomers, and in the end it may cause more harm than good. You do not need to use all of these advanced features. You can also restrict the language features, selecting a subset that seems simple enough that it will avoid any undesirable complexity. This is what John Carmack did when writing *Doom 3*. He simplified a complicated language (C++) and kept only the most useful functions, which ultimately helped to make the design simple, which made it easy to understand and maintain by other developers.

Notes

1. The program has been compiled with the -O0 flag to avoid any side effect of compiler optimization.

2. http://valgrind.org/.

3. https://dwheeler.com/flawfinder/.

4. https://gcc.gnu.org/onlinedocs/gnat-style.pdf.

5. https://gitlab.com/pycqa/flake8.

6. https://www.python.org/dev/peps/pep-0008/.

7. The full text of this and other Voice of the Practitioner sections can be found in the appendix under the relevant name.

8. https://www.code-inspector.com; https://www.codeclimate.com/.

Further Reading

A 2007 blog post written by Jeff Atwood (which can be found at https://blog.codinghorror.com/twitter-service-vs-platform/) detailed Twitter's early problems with scalability and Ruby on Rails.

A good example of feature restrictions is MISRA-C, a subset of C for embedded, safety-critical software. Another good example is the legendary programmer John Carmack that, while implementing the ID4 game engine, reduced C++ to C with classes to avoid any C++ traps and pitfalls and facilitate writing code.

Security issues are probably the bugs and debt items that people are most interested in because they can have huge impacts on a company or on product trust. There has been a lot of effort to make code more safe and secure. The SEI CERT C coding standard and its related books (Robert C. Seacord, *Secure Coding in C and C++* [Addison-Wesley, 2013]) is a good start to learn how to write secure code. There are variations of this book for other languages (such as Java). Marc Espie, developer of the OpenBSD project (one of the most secure operating systems) produces a very good class on secure programming (Security Development 101: https://www.lse.epita.fr/teaching/epita/sede/slides-sede.pdf). The references for Common Vulnerabilities and Exposures and Common Weakness Enumeration are the websites hosted by MITRE: https://cve.mitre.org/ and https://cwe.mitre.org/). The Heartbleed bug (http://heartbleed.com/) is one of the most harmful vulnerabilities recorded to date.

There are also several good catalogs of security patterns, such as Ramesh Nagappan and Christopher Steel, *Core Security Patterns: Best Practices and Strategies for J2EE, Web Services and Identity Management* (Prentice Hall, 2005). Finally, the article H. Cervantes, R. Kazman, J. Ryoo, D. Choi, and D. Jang, "Architectural Approaches to Security: Four Case Studies," *IEEE Computer*, November 2016, 60–67, is the source of the security framework box used in this chapter.

If you are looking to have insights about what programming languages or tools developers are using, StackOverflow does an annual survey in its community to have metrics on what tools, methods or languages are used. You can access the survey report on https://insights.stackoverflow.com.

Robert C. Martin, *The Clean Coder: A Code of Conduct for Professional Programmers* (Addison-Wesley, 2011) is a reference regarding coding guidelines and best practices to keep your code free of anything that might incur technical debt.

The DRY principles have been well-documented over many years. Some useful references include Steven Foote, *Learning to Program* (Addison-Wesley, 2014); and Andrew Hunt and Thomas David, *The Pragmatic Programmer: From Journeyman to Master* (Addison-Wesley, 1999).

The claim that software maintenance takes between 50% and 80% of total project cost has been around for several decades. While the precise numbers vary, all of the research agrees that the percentage of a system's total cost that goes into maintenance is within this range, and most put it at the high end of the range (i.e., closer to 80%). See Pierre Borque and Richard Fairley, eds, *SWEBOK—Guide to the Software Engineering Body of Knowledge, Version 3.0* (IEEE Computer Society Press, 2014), for a comprehensive discussion of maintenance, its types, and its costs.

How to avoid a lot of design issues and code duplication? By using and reusing design patterns! Design patterns were introduced by Erich Gamma, Richard Helm, Ralph Johnson, and John Vlissides, *Design Patterns: Elements of Reusable Object-Oriented Software* (Addison-Wesley, 1994). Another great resource is Eric Freeman, Bert Bates, Kathy Sierra, and Elisabeth Robson, *Head First Design Patterns: A Brain-Friendly Guide* (O'Reilly, 2004) that explains design patterns using Java in very simple terms.

Cory J. Kapser and Michael Godfrey, "'Cloning Considered Harmful' Considered Harmful: Patterns of Cloning in Software," *Empirical Software Engineering*, December 2008, gives another perspective on code duplication, especially regarding potential experimentation and exploration.

If you want to look at a good example of coding style and coding guidelines, there are some very good examples online. The Linux coding style (found at https://www.kernel.org/doc/html/v4.10/process/coding-style.html) or NetBSD coding style (found at ftp://ftp.netbsd.org/pub/NetBSD/NetBSD-current/src/share/misc/style) are very good examples to follow. You can also have a look at the *Doom 3* coding style (found at https://fabiensanglard.net/fd_proxy/doom3/CodeStyleConventions.pdf), which is a great example for a C++ codebase.

If you love to read code and explore a large codebase in detail, we highly recommend the website of Fabien Sanglard: http://fabiensanglard.net/. Fabien's website contains code reviews for very large codebases, such as Git or the most popular ID software games (*Doom*, *Quake*, *Wolfenstein 3D*). Fabien not only explains how programmers kept the code clean but also how they had to make trade-offs to deliver the products they wanted (e.g., some nice optimizations to speed up rendering in a 3D engine that will unfortunately make the code less readable or maintainable). This website is a goldmine for engineers willing to spend some time exploring clean and state-of-the-art code bases.

6 Testing Debt

> There's nothing that can compare to testing yourself the way you do every time you step in the ring.
>
> —Sugar Ray Leonard

> Testing leads to failure, and failure leads to understanding.
>
> —Burt Rutan

Testing software is important. Tests are an important way to reveal potential technical debt and software bugs. However, test code can also contain technical debt. In this chapter we discuss how to identify, manage, and avoid testing debt.

Too often, developers ship code without appropriate testing. There are many reasons for this, but the main one is the pressure to deliver the product on time. At first, this strategy pays off, especially if you are in the early stages of your development—such as in a prototyping or a rapid iteration context: you are able to iterate fast and ship new features quickly. Developers can fix broken code in a few minutes (or hours), and at that time it doesn't really matter if you break something that is going to be fixed quickly. As your product grows over time, your team changes, developers leave, new ones join the development team, and new features are piled on top of existing code. Fixing any buggy code starts to take days or weeks. There may be broken features nobody knows how to fix. All the agility you had in the beginning of the project vanishes in a very short time.

You may find this story very familiar and yet, it repeats itself all the time in the software industry. It is common to hear developers, working on large established codebases, complain that fixing code and flaky tests can take days or even weeks.

Insufficient testing affects large companies, too. Windows 95 was hanging after forty-nine days of usage because it counted the time in milliseconds and the system blocked because of an integer overflow. In 2015, it was found that the Boeing 787 had to be rebooted every 248 days because of a similar problem. You would think that the aviation industry would have learned the lesson, yet in 2019, the Airbus company mandated its A350 plane be rebooted every 149 hours for a similar problem. Were these issues design issues, or would they not have existed with more testing?

In this chapter, we discuss good testing practices and explain how testing relates to technical debt. While test code itself may have technical debt, the biggest problem is that inadequate testing is itself technical debt. Having bad testing practices will ultimately incur more technical debt. We first discuss why testing relates to technical debt. We then explain how we can identify a lack of testing, how we can manage this lack of testing, and how we can avoid testing problems in the future.

6.1 How Testing Relates to Technical Debt

Testing provides a way to verify that your code and system behave "correctly," according to some defined behavior (typically, your specifications or requirements). Testing can be done at different levels, from fine-grained tests of specific methods or lines of code, to system-wide product testing, as we discuss later in the chapter. The main benefit of testing is to verify that, given an input, you see the expected behavior or value. Technical debt in tests come from shortcuts taken in the testing process, which leads to inadequate or uninformative tests. The goal of testing is (or should be) to cover the entire state space of the program. While this is virtually never achieved in practice, that should be our goal. A number of complexity metrics have been devised to attempt to measure the complexity of a code base, and hence the complexity of the test suite needed to cover that code base. The most famous of these is McCabe, or Cyclomatic complexity. Cyclomatic complexity measures the number of independent paths through a program's source code. For example:

```
if (condition == True)
// do something
else
// do something else>
```

The idea is that if you have a snippet of code, then you should have (at least) two test cases, one that exercises the path where condition == True and one where condition == False.

Given this perspective, we can identify several causes of technical debt in testing:

- **Inadequate testing,** with key functionality or even modules lacking tests. Clearly if we want to cover the entire state space of a system, the very minimum testing that we can do is to ensure that every line of code in every module is executed.
- **Testing the wrong things,** that is, testing methods with little potential for impact. Given that we can almost never completely test a non-trivial program, we should prioritize our testing efforts on the areas of highest value and of greatest risk (if they fail).
- **Needless testing,** testing parts of the system you already know are correct (e.g., old code that is reliable). Again, if testing effort is limited (and it is always limited) then you want to spend your time on the areas of the system that are likely to reveal a previously unknown bug.
- **Tests that are unreliable and flaky,** with low signal-to-noise ratios.
- **Testing only the happy path** (e.g., success cases and not testing error cases).
- **Tests that don't match the specification or documentation.**

Let's take the example of an authentication system: you should expect that the system authorizes you when you provide the correct username and password combination and rejects you when you provide an invalid one. Here, the inputs are clear and unambiguous (username, password) as is the expected behavior (login success or failure). Tests should be deterministic, to the extent possible (e.g., given the vagaries of distributed systems). Determinism means that the test result should always be the same and should not depend on any environmental factor (such as the current time—good testing frameworks have functions to handle contextual factors like time and date).

Testing ensures that:

1. **The actual code works:** it validates that what is being shipped actually does what it is supposed to do. This is useful in the present, when you are developing the code.

2. **Future changes do not break existing functions:** having tests in place (that are run periodically, at every code change) guarantees that new code changes do not silently break what is already implemented.

6.2 The Increasing Recognition of Testing

Testing has received more attention and gotten more traction over the last two decades. Until the 1990s, software testing was primarily manual: developers executed the program themselves or a quality assurance (QA) team followed a script to manually test the product. This process is expensive (you need to hire engineers to test), unreliable (you need to trust humans, who are making errors by definition), and incomplete (manual testing is hard to scale to completely cover a system). With the increasing adoption of continuous integration and continuous deployment approaches over the past decade, the software development process is increasingly making use of automated testing, which has the following benefits. First, automation makes it easier to test: with automation, testing just happens, and becomes as much a part of workflow as compiling. Second, automation ensures repeatability, so that everyone's changes are running (mostly) the same set of tests. In combination these benefits increase the quality of the project (as long as appropriate and sufficient numbers of tests are written!).

Developers and managers today recognize the importance of testing in the software industry. There has been an increasing focus on testing with the development of numerous libraries and products—for example, the ability to mock objects, the xUnit testing frameworks,[1] and more recently, the increasing number of user interface testing tools for mobile applications and websites.

6.3 Identifying the Lack of and Need for Testing

Technical debt in testing reveals itself most commonly through downstream interest costs. In particular, we will see increasing bug reports from released software. This means the tests we do have are either inadequate or not testing the proper parts of the software product.

6.3.1 Loss of Agility and Increase in Bugs

The first symptom of a lack of testing is an increasing number of bugs (e.g., new features regularly break existing ones) and a loss of agility (the time to deliver fixes or new features). Without appropriate testing, broken code will inevitably be shipped with bugs, with some features not working properly. It will also take longer to find and fix bugs. Both things invariably increase the cost of delivering software.

Developers usually don't notice these symptoms at first. Often, product managers get feedback about bugs and broken functionality from users, and they then forward this feedback to the software managers and developers. This long feedback loop increases both the time to fix the bug and customer frustration. As there are no adequate tests in place, it takes more time for developers to find the exact cause of the bug, thus reducing development velocity and agility.

As mentioned before, the biggest return on the investment of maintaining an extensive test suite is to avoid bugs as much as possible. This is where lack of testing incurs technical debt: sooner or later a product evolves, its code changes, and without appropriate testing there is the near certainty that existing functions will break. Discovering what broke, what caused it, and how to fix it will incur a significant amount of work, whereas a test could have simply pinpointed the culprit at lower cost and without exposing this dirty laundry to customers. Testing provides the confidence to make frequent changes.

6.3.2 Levels of Testing

Testing takes place at different levels of granularity, from individual functions/objects to the complete integrated system. Depending on the source, you might find different names for each level, but the names are not what matters; what actually matters is to understand the different levels of granularity and what you need to do to test at each level. Technical debt can be introduced if you short-circuit testing in any level of granularity.

Unit tests Unit testing consists of checking an atomic unit of code. In concrete terms, it means that you test an atomic unit of functionality in your code, which may be a function, a method, or even a statement (e.g., to do data access). The type of unit you test depends on the programming paradigm and importance of verifying that unit's behavior. When doing unit testing, the behavior of external services are simulated or mocked. This is the lowest-level of testing one can have and ideally will exhaustively test the success and failure cases of code units. By "exhaustive" we mean that the tests should be sufficient to give you confidence in the unit's behavior. A simple example would be testing of a square root function. Exhaustive testing does not mean "every real number" but rather the space of important input classes, such as negative numbers, zero, positive numbers, floating point values, and so on.

Technical debt at the unit test level mainly comes from inadequate testing, tests that fail to conform to the spec, and tests that are flaky.

Integration tests Integration testing consists of verifying that units of code interact with each other according to the specifications. It consists of connecting different units of code together and checking that they behave according to the specifications. The importance of such tests is often overlooked, which is unfortunate because this is where most errors come from.

In particular, when units of code are developed by different independent contractors, integration testing ensures that requirements have been correctly implemented by all parts and that inputs and outputs are consistent. While this seems simple and perhaps even obvious, skipping integration tests can lead to disasters, such as the one from the Mars Climate Orbiter. In that example, units were expressed in both standard and imperial notation (e.g., m/s and ft/s), resulting in a missed orbit. Such tests should be automated (i.e., executed at each code change). However, depending on the development process (e.g., modules being developed by different teams or suppliers), this is not always possible and thus would require manual integration.

Technical debt in integration tests is due mainly to inadequate testing and happy path testing. It is easy, but insufficient, to test the integration result that should happen. Integration tests should also examine what happens when things go wrong. It is at the edges of module boundaries that many larger errors can occur.

Functional tests Functional testing consists in confirming the correct behavior of a component without considering implementation details. This is often referred to as black box testing. The benefit of functional testing is that you can test different implementations of the same interface without having to change your tests.

For example, if you test a website, you don't test the actual code of the system, you simply send requests and check the responses (accuracy of the response), their correctness (is the HTTP response code 200, 4xx or something else), and typically some other factors (such as latency or throughput). You actually don't need to know how the server is implemented to perform your tests and you can reuse the same tests on different platforms.

Functional testing is extremely important when you:

1. Compare two competing products (such as the ACID tests that check standards compliance of web browsers).
2. Evaluate different implementation alternatives based on some obvious factors (such as latency, CPU utilization, etc.).
3. Refactor or rewrite a software component from scratch. Functional tests will then check that the new product behaves like the old one and does not introduce regression bugs.

Technical debt in black-box testing is incurred when the spec and the black-box test are misaligned. It is also easy to find flaky tests since we have little insight into the component's behavior. Prioritization is vital at the functional level, since the input space is potentially large.

6.3.3 Appropriate Levels of Testing

The level of testing you should select depends on (1) the maturity of your product and (2) the domain of your product. If your product is immature, the change velocity is typically high, and therefore tests must be constantly updated as well. For that reason, it might be wise to only have unit testing as a first step, and when the product matures and the codebase stabilizes, start doing some integration and functional testing. If your product is mature and has had customers for several years, your focus is no longer on developing new features every day but rather making sure that the product stays stable and doesn't have any regression bugs (bugs that get introduced due to new features conflicting with old features). In that case, you should put effort into improving testing and including integration and functional testing.

There is one exception to this rule that applies to the safety-critical systems domain (such as aerospace, avionics, military, or self-driving vehicles): no matter the development stage, because your system is life- or mission-critical, you need to have extensive testing from day 1. When releasing any version of your system, you will not only need unit and integration testing but also functional and final acceptance testing. If you are working in these domains, it is wise to invest massively to automate testing. This will make it more efficient so that you can reduce testing time and effort. In many cases in this domain, testing is done by an independent third party to make sure that requirements are correctly implemented and tested by both parties (a process that often discovers inconsistent requirements at the same time) and that developers not only test the happy code path but also all of the edge cases. The happy path is what you hope your code is doing, when there are no defects or bizarre conditions that take less frequent code paths.

6.4 Managing Your Testing Activities

6.4.1 What Can Be Tested?

Today, almost everything can be tested. The most difficult thing to test is probably graphical user interfaces. The code that interacts with the filesystem, network, and popular microservices can be tested without many problems. Such services can be simulated and no matter what languages you use, there will almost always be a good library to mock existing services.

On the other hand, testing user interfaces is challenging:

- All the configuration combinations cannot be tested (resolution size, web browser in the case of a web interface, configuration—Linux, Windows, Mac, etc.).
- As the UI displays pixels on a screen, testing will verify values of pixels (e.g., the color pixel at position X, Y is white after performing a specific series of operations), which is also very fragile.

There are some frameworks (most are proprietary) that simulate user activity and allow developers to programmatically trigger user activity (mouse, keyboard, touchscreen) and check the user interface status. Most of the time, such frameworks are so difficult to set up and so hard to use that many development shops are still testing graphical user interfaces manually.

Lastly, while almost everything can be tested, the hardest part is to have an extensive testing suite that includes failure conditions and potential outages from external systems. For example, when developers design and implement a system that relies on Amazon S3, a data storage service that is known to be very robust, they will not likely test their system against a potential S3 outage, assuming this system never fails (but this in fact happened in 2017). While this is testable by mocking the S3 interface and constantly returning an error or timeout, it is easy to overlook some rare conditions. This is why it is crucial to have extensive testing and to ask for more tests when reviewing code (see chapter 5 on effective code reviews).

6.4.2 Measure Code Coverage

Code coverage analysis is used to trace what code is being executed by your tests. In other words, it is a measure to report on what parts of your code are currently tested, showing you how confident you can be about your test suite. It is a rough proxy for test adequacy (but a high coverage number does not guarantee you have adequate testing).

Code coverage is used to:

1. Report that code paths are being executed during testing
2. Detect potential dead code and gaps in testing (what parts of the code have not been tested)

There are several levels of test coverage (statement, decision, modified condition/decision coverage) and unless you are working on a safety-critical systems (where coverage methods are dictated by safety standards), you should target > 75% statement coverage. That will ensure that most of the code is being executed during your tests, giving you reasonable confidence that (1) a large portion of the code is being tested, and (2) this large portion of code behaves as you expect.

Today, almost all popular languages have tool support to measure code coverage. To illustrate how to practically test your program coverage, we will show how we can use a code coverage tool. We will use the *coverage* tool for Python, a very popular code coverage tool.

Let's consider a simple program that parses a `passwd` file (located in `/etc/passwd` on most UNIX systems) as a concrete example. The following Python code reads the file and stores information for each user in a dictionary. The program handles several edges cases: when the file is invalid, when a user was already processed, and if the content itself is invalid. This is not a great program but no programming crime has been committed here.

```
import os

class InvalidFileException(Exception):
    pass

class InvalidFileFormat(Exception):
    pass

def read_passwd(filename):
    if filename is None:
        raise InvalidFileException("invalid argument")

    if not os.path.isfile(filename):
        raise InvalidFileException("file does not exist")
    res={}
    with open(filename) as f:
        for line in f:
            parts=line.split(":")
            if len(parts) >= 3:
                if parts[0] in res.keys():
                    raise InvalidFileFormat("username {0}
                    appears
more than once!".format(parts[0]))
                elif not parts[2].isdigit():
                    raise InvalidFileFormat("uid {0} is not a
                    number!".format(parts[2]))
                else:
                    res[parts[0]]=int(parts[2])
            else:
                raise InvalidFileFormat("Non conformant line
                {0}".format(line))
    return res

if __name__ == "__main__":

    read_passwd("passwd-test")
```

Let's see now how we would test such a program. We use the unittest Python library and define a test for each particular condition: when there is no file (`test_no_file`), when the filename is invalid (`test_invalid_file_name`), and when everything goes right and the function returns something (`test_file_ok`). Note that there are more tests for the failure cases than for the happy code path case.

```
import unittest
from lib.passwd import read_passwd, InvalidFileFormat,
InvalidFileException

class TestCase(unittest.TestCase):
    def setUp(self):
        pass
    def test_no_file(self):
        with self.assertRaises(InvalidFileException):
            read_passwd(None)
    def test_invalid_file_name(self):
        with self.assertRaises(InvalidFileException):
            read_passwd(None)
        with self.assertRaises(InvalidFileException):
            read_passwd("/tmp/foo/bar/")
    def test_file_ok(self):
        res=read_passwd("test/data/passwd")
        assert(res["julien"] == 1000)
```

Once the tests are written, the coverage tool is used to measure the code coverage from the test execution (as shown below): it gives a good overview of what files are being covered by the tests and how many statements are not being tested, letting developers know what additional tests to write to improve coverage (other output exists, showing exactly the lines not being covered). Note that the code coverage is not 100%, since the tests we wrote did not include tests that check the program with invalid content (when the `InvalidFileFormat` exception is raised)

```
$coverage report
Name                Stmts   Miss   Cover
--------------------
lib/__init__.py       0       0    100%
lib/passwd.py        22       3     86%
--------------------
TOTAL                22       3     86%
```

6.5 Avoiding Testing Debt

Testing debt accumulates when we do inadequate testing and take shortcuts like only testing our happy paths, or commit only to testing features and not system qualities (see box 6.1). So how do we avoid such debt? That is, how do we ensure that we do adequate testing from the get-go? We now tackle this issue.

6.5.1 Adopt Test-Driven Development

We hope you are now convinced of the usefulness of extremely thorough testing! Unfortunately, writing tests *after* writing the code is often not enough. In fact, the benefits of testing can be reduced because of the way we wrote the tests:

1. **Developers test only success cases** (the happy cases) and not failure conditions, resulting in having tests that do not cover all requirements, especially failure cases or edge cases.
2. **Developers tend to modify tests** based on what is implemented rather than fixing the software itself. For example, if the software has a defect and reports an incorrect value (for example, returning null for an error instead of raising an exception), it is not uncommon to see a test checking the erroneous behavior (having a test checking for null rather than checking for an exception). Writing a test that does not check the specification is a form of technical debt because the tests no longer reflect the original requirements.

As a disciplined approach to testing is becoming more prevalent in the software industry, developers are starting to embrace the test-driven development (TDD) approach. When using TDD, developers write the tests before writing the software. You capture your software specification (for example: *my API should return the list of JSON elements in that specific order*) as tests before you even write the code that implements these requirements. If your system depends on external services (such as an external API or a data repository), they are mocked (e.g., simulated). In the TDD paradigm, only after you have written all tests for that unit of functionality do you start writing code. However, this is still important to ensure you have good code coverage, as using TDD does not guarantee good code coverage.

There is a huge benefit to this approach: you have a clear idea, *before* writing any piece of code, of the different success and failure conditions. It helps you to consider and handle all edge cases, and this increases your confidence that your implementation actually works. This is a breakthrough change compared to prior development approaches, where developers wrote the code first and then wrote the tests only when the code was working (and, as we said, they often only tested the happy code paths, omitting most if not all edge cases). TDD is also a useful way of addressing the need to create a culture of testing. TDD practices enforce the importance of testing as a crucial part of the software development process.

Your software must satisfy all of its requirements, and TDD is a fundamental process to help guarantee that your requirements are being correctly implemented in your code. This process is definitively time-consuming up front. But it is one of the best weapons you have to protect yourself against potential bugs that could be introduced down the road.

6.5.2 Maintain and Analyze Test Runs

As with software metrics (which we described in chapter 5), it is important to implement testing metrics and make sure that they are improving over time. The key metrics you should focus on here are:

1. **Number of tests:** How many tests are being written? Do not track only the number of tests but the relative number of tests compared to the unit of code you are testing. In general, each unit (function/object) requires a few tests (at least one for the happy code path and several others—one for each failure case).
2. **Test coverage:** keeping a history of the test coverage helps you to check that it is improving over time and does not go below a critical threshold.

Box 6.1
Voice of the Practitioner: Andriy Shapochka

> [On one project], as a rule unit, integration, and load testing are implemented to some degree, but since project sponsors are reluctant to commit resources to tests which obviously do not directly add up to the features of the implemented system and are not understood by the decision makers, the tests are frequently insufficient, out of date, or do not bring much value, being too simplistic to properly test business logic, complex scenarios, extreme loads.
>
> —AS[2]

As mentioned earlier, a good coverage measure is more than 70% or 80%, but it is important to not only focus on quantity but also quality. A high coverage measure alone is not a measure of success (developers can omit writing tests that exercise failure cases, thereby increasing the code coverage without actually increasing the quality of the tests). It is important to make sure your tests cover the failure cases as much as success cases (tests that cover both success and error paths should be required during the code review).

There are several services that allow you to run and manage test coverage analysis, and these services integrate with configuration management systems. Such tools should be automatically integrated with your build system, should execute all tests, maintain historical metrics, and show you how these metrics evolve over time. Figure 6.1 is a screenshot of such a service, surfacing test coverage information (such as coverage per file, number of lines covered/missed, etc.).

6.5.3 Automate Testing Activities

Let's now assume that you have adequate test coverage. That still is not enough. Test execution must be automated in your continuous integration pipeline and triggered either:

1. **When a code change is submitted for review:** by executing the tests on each piece of code being submitted for review, the authors are notified of test failures and can immediately see potential regressions they introduced (or the need to modify the test if the system specifications changed).

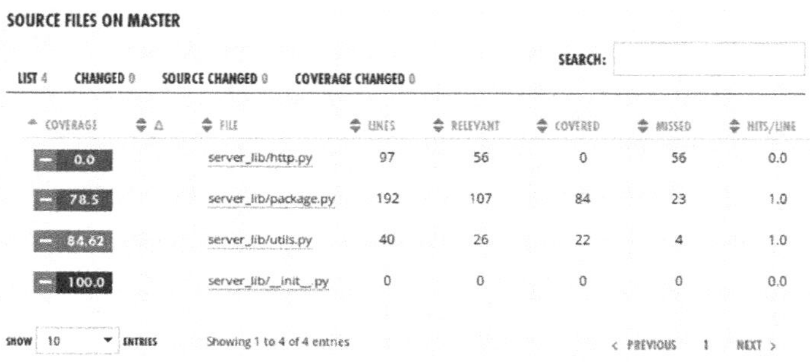

SOURCE FILES ON MASTER

SEARCH:

LIST 4 CHANGED 0 SOURCE CHANGED 0 COVERAGE CHANGED 0

COVERAGE	Δ	FILE	LINES	RELEVANT	COVERED	MISSED	HITS/LINE
0.0		server_lib/http.py	97	56	0	56	0.0
78.5		server_lib/package.py	192	107	84	23	1.0
84.62		server_lib/utils.py	40	26	22	4	1.0
100.0		server_lib/__init__.py	0	0	0	0	0.0

SHOW 10 ENTRIES Showing 1 to 4 of 4 entries < PREVIOUS 1 NEXT >

Figure 6.1
Example dashboard offered by coveralls.io: coverage metrics are showing which files have appropriate levels of testing and which files require more testing.

2. **When a code change lands in your production (e.g., master) branch:** even if you execute tests when new code is being submitted for review, you cannot guarantee that all tests will pass when the code change is merged into the codebase. After all, another developer could have made another change that breaks this test). For that reason, all tests should be re-executed when a code change lands **and the change should be rejected if any test fails.** An invariant in your repository is that all tests in the production branch should pass (and any failed test should be considered as the highest priority for fixing).

Such automation can be implemented by installing hooks in your configuration management system (e.g., Git, Subversion). Popular hosting platforms (GitHub, GitLab) offer options to set that up in few clicks. While this will typically cost a few dollars a month, it is a good and inexpensive investment considering the cost of a software engineer's time and the cost of a mistake.

6.5.4 Avoid Manual Testing

Some people still test their software manually: they deploy a product, use it, and test that everything is working. Unfortunately, manual testing has a lot of limitations:

1. **Scaling:** you cannot afford to run tests for each code change (or you will need an army of testers working for you 24/7).
2. **Reproducibility:** if you do your own testing, then the product is tested on a specific environment (yours) and not a standardized environment. This can introduce variability in the compilation or deployment and can introduce unexpected bugs that are hard to find and reproduce.
3. **Completeness:** it is extremely hard to test all conditions and edge cases for a program. For example, if you test a service connected through the network, you might want to test it using different bandwidths; if you test a UI, you want to test it with different screen sizes or devices. These test parameters may be difficult to reproduce on your own but can be simulated.
4. **Error-prone:** manual testing is a human activity that, by default, makes it error-prone. For example, one person may interpret a requirement differently than another person, introducing uncontrolled variability.

While tests can be automated the vast majority of the time, there are unfortunately some cases where manual testing is still required, especially for integration testing of embedded devices, where functional, end-to-end tests should be done. Manual testing should, however, be the exception, and not the norm.

6.6 Conclusion

Testing is one of your best friends in your fight against technical debt. This chapter introduced best testing practices to reduce your technical debt. Testing checks that your application behaves as expected but, perhaps more important, ensures that you do not introduce any regression bugs as your product evolves. For example, if you introduce technical debt in your testing, if your testing strategy is weak (e.g., code coverage is low or nonexistent), or if you do not monitor your level of testing, you need to start to build the necessary test suite and support infrastructure right away! The goal is not to target full (100%) coverage but to have enough tests and coverage to give you confidence that the most critical problems will be caught during testing in an efficient, fully-automated way. Today there are enough tools and platforms to help developers in their testing efforts. We encourage developers and managers to start using them to avoid technical debt.

Notes

1. https://en.wikipedia.org/wiki/XUnit.

2. The full text of this and other Voice of the Practitioner sections can be found in the Appendix under the relevant name.

Further Reading

We recommend reading about testing. In particular we recommend *Introduction to Software Testing* by Jeff Offutt and Paul Ammann (Cambridge University Press, 2016). If you want to learn about test-driven development, we recommend Kent Beck, *Test-Driven Development by Example* (Addison-Wesley, 2002). Kent Beck is also the author of the JUnit testing framework (among other things) and a great innovator in the domain of testing and, more globally, software engineering. Almost any publication by Beck would help you to understand concepts related to testing.

The earliest test metrics, such as McCabe Complexity (T. McCabe, "A Complexity Measure," *IEEE Transactions on Software Engineering* 4 (December 1976): 308–320), make for interesting reading and might be useful as a gross guide to the amount of testing effort you should commit to, but McCabe Complexity has not been found to be significantly better than lines of code, at least as a predictor of defects.

If you are not familiar with testing frameworks, we strongly encourage you to have a look at the most popular testing frameworks in the industry (depending on your programming language). For Java, the book *JUnit a Cook's Tour* from Erich Gamma and Kent Beck is a good introduction (http://junit.sourceforge.net/doc/cookstour/cookstour.htm).

If you want to read more about the different levels of code coverage (and especially if you work in a safety-critical domain), you might want to read the following article: Chilenski, John Joseph, and Steven P. Miller, "Applicability of Modified Condition/Decision Coverage to Software Testing," *Software Engineering Journal* 9, no. 5 (1994): 193–200. It explains modified condition/decision coverage in detail, a code coverage analysis technique that was introduced in 1994 to qualify avionics software at the highest criticality level. While it might not apply to many projects, it gives a great overview of the different levels of code coverage techniques.

There are multiple frameworks and tools to test your user interface. The most popular is Selenium and the number of frameworks to validate your user interface grew as more web applications were being developed. Open-source tools such as Appium are very powerful, well-integrated with existing frameworks, and have good support.

The story of the Mars Climate Orbiter is dissected in the official (and readable) NASA report, available at: https://llis.nasa.gov/llis_lib/pdf/1009464main1_0641-mr.pdf. A summary of the Amazon S3 Service Disruption in the Northern Virginia Region is at: https://aws.amazon.com/message/41926/. A web search for "post-mortem AWS" (or other services) will turn up more such reports. In general system engineers are good at reporting on failures.

To understand how we can measure the value of testing—that is, the cost-benefit ratio of writing new tests vs. the value of the information those tests provide, you can look at the report from Ericsson: Armin Najafi, Weiyi Shang, and Peter C. Rigby, "Improving Test Effectiveness Using Test Executions History: An Industrial Experience Report," in the *Proceedings of the 41st International Conference on Software Engineering: Software Engineering in Practice* (2019).

Case Study B: Twitter

Summary and Key Insights

In its early days, Twitter faced many technical challenges for scaling its services. The service was often inaccessible, and the only page displayed was the infamous fail-whale, a page that was shown when the service was unavailable. Today, Twitter is one of the most robust services in the world, relaying thousands of tweets per second on users' timelines. How did engineers turn a system that scaled poorly into one of the most reliable systems? How were errors diagnosed and fixed without service interruption?

This chapter explains how technical debt played a key role in the problems Twitter faced and how Twitter engineers changed the narrative and transformed a website with repeated stability issues into one of the most stable websites on the planet. This is a focus on technical debt as it was incurred, identified, managed, and eventually repaid. It details the problems faced by the company, and how they fixed them over several years. In particular, we show how the use of large-scale design changes, including a shift to microservices and new frameworks, and a singular testing focus, helped to reduce the debt burden.

This case study is the result of interviews with past Twitter employee Kevin Lingerfelt and the publicly available documents about the company.

Background

Twitter started as a Ruby on Rails website, developed by only a few engineers. After only a few months, the popularity of the service exploded, and Twitter faced many growing pains. The technology selected (Ruby on Rails with MySQL) did not scale well and often the website was down.

Twitter was initially a monolithic, giant app that had to be redeployed anytime a code or configuration change was done. The technology stack (Ruby on Rails and MySQL) did not scale with user demands, causing the fail whale (an image that signified a failure of the system) to appear. For a long time, Twitter fell into the technical debt trap. When we interviewed Kevin, he defined technical debt: *"It is just like other kinds of debt in that it provides leverage: you can achieve something quickly, but you have to pay it back later. It fits with the financial metaphor: if you don't pay it back later, it will create real problems."* Twitter was mired in technical debt mud for some time. And yet, with hard work, they managed to get out of it. The next sections detail the problems Twitter faced and how they address them.

Identified Problems

A Custom Technical Stack

As Twitter was facing scaling issues, engineers decided to fork major components (such as the Ruby runtime and its web framework Ruby on Rails) and modify them locally to fix some scaling issues. For Ruby, this resulted in a new fork, Kiji, that had to be maintained separately for the main branch. The fork gave immediate short-term results (e.g., better garbage collection and improved performance) but dragged the company into a maintenance nightmare. At first, the fork was not backported into the main Ruby code. Major changes from Ruby then had to be backported (or not), costing development time and putting the stability of the Kiji runtime at stake.

More importantly, security patches from upstream versions of Ruby had to be backported into Kiji, and Twitter engineers had to closely monitor what vulnerabilities were potentially affecting Kiji. The same issues also apply to the local fork of Ruby on Rails: the versions forked by Twitter (2.0.2) were no longer supported by the community, and security patches were only released for new versions. Twitter engineers not only had to figure out if a security patch applied to their fork of Ruby on Rails but also how to apply the patch. This was time consuming and also very error-prone, as the patch was applied manually.

The technical debt metaphor clearly applies here: the decision to fork brought short-term gains (better performance and slightly better ability to scale came immediately) but affected the company over the long run because the debt was not paid down (e.g., when changes were not backported into the upstream branches of Ruby, Ruby on Rails, and other technologies).

Lack of Agility and Reliability

Twitter started like most other startups, with a simple, single codebase. But the company grew quickly. So did its codebase, its features, and the size of its workforce. This raised several issues. At first, the codebase was changing very quickly. This created problems for developers: a change that was working on a local branch might be broken few hours later, because somebody else has changed the code. It was really hard to keep up with the pace of changes on the giant codebase.

Such issues could, in theory, have been caught by testing, but code coverage was also low at that time. As a result, bugs were discovered at run-time, and often by end-users. For the backend, this translated into incidents that could crash the website. For the frontend, this could lead to the website not even rendering and leaving the user with just a blank page. Twitter engineers gave a name for this kind of issue: the white screen of death (WSOD—an analogy with the blue screen of death, or BSOD, seen on Windows).

Such crashes required rollback (redeploy the previous stable version) and discovering what caused the issue after the fact. Developers were dissecting the source code history, trying to find the line of code responsible for the bug of the day or week. Deployment became unpredictable to the point that the company had meetings where managers would have to explain what was being shipped so that system reliability engineers (SRE) were aware of the changes and potential operational impacts. Teams were not operating at their own pace, and deployment time was dictated by the frequency of these meetings.

Needless to say, all of these factors contributed to decrease agility: changes took longer to deploy, and the site had a lot of reliability issues. It took developer time; developers could have spent time developing new features but instead spent time dissecting the commit history and adapting their code so that their new features could be deployed. Impact on productivity was real.

Scalability

Ruby is an interpreted language, which has great benefits, especially when you want to iterate fast, but it also has significant issues. In particular, errors are surfaced at run-time, causing reliability problems, as discussed previously. Also, interpreted languages are very convenient for prototyping and getting to a minimum viable product fast, but they are not as fast and performant as other solutions (such as Java, Scala, or Kotlin that are executed on top of the Java Virtual Machine). And such languages create well-known scalability issues. In the case of Twitter, these issues were clear: the application was single-threaded and was consuming 1GB of RAM. Having a giant application implemented with an interpreted language involved a tradeoff. This simple model made it very easy to provision capacity: one active request is being served by one instance and we know how many instances can be run on a machine based on the RAM capacity, but it was then very expensive to scale.

There was no way to decouple different services and deploy more instances of one particular service: everything had to be deployed in this one, giant application. This led to an enormous waste of resources and made it difficult to scale. And it was impossible to deploy more instances of the program that was the most used.

Complexity

As performance and scalability were major issues, engineers were working hard to mitigate them. To reduce load on the backend, engineers started to add caches at different levels of the main application. These caches provided some short-term improvements in terms of performance but introduced a lot of complexity that had to be managed later when modifying or updating the application.

This complexity became an issue for each new hire or person not familiar with the code. There were multiple levels of caching and a change on each of them might have unexpected consequences. This obviously decreased agility and it quickly became a priority to simplify how data was cached.

Solutions

From the problems mentioned earlier, it was clear that taking on more debt would eventually lead to technical bankruptcy. The company recognized this and started an initiative to address all of these problems. It took more than six years to complete this switch (the first services were fast to migrate but the remaining services took three years to migrate) and to arrive at the levels of performance and stability that all users experience today.

Micro-Services to the Rescue

The first obvious move was to rearchitect Twitter as a whole. Having all services (tweets, users, ads, etc.) in a single application was the source of many problems. This monolith was refactored into multiple small services. This new architecture led to immediate gains:

- **Agility:** teams can change and deploy code at their own pace instead of being dependent on other people's code (and on the big meeting that decides what can be deployed and when).
- **Reliability:** a crash in one service can no longer affect another service, therefore reducing the impact of a change and the number of potential outages.
- **Complexity:** as the code base was simple and easier to understand onboarding became more simple, and new employees could ramp up more quickly.
- **Scalability:** as services were separated and isolated, it became simple to start more services in areas with more requests. For example, it was then possible to run more instances of a service that serves your timeline than services that modify a user profile (assuming that users are more likely to read their timeline than modify their profile).

It is clear that adopting a microservice architecture was a major win for Twitter from several perspectives. While many projects start as a monolith (because it is easier to understand, modify, and update for small teams), rearchitecting is important to do when scaling. This case study exemplifies the issues that can arise when the debt is repaid too late.

However, this change was not the only one required to update the tech stack and make Twitter more reliable. The next section explains what changes have been made to the software stack to make Twitter's software faster.

Adoption of Scala and the Java Virtual Machine (JVM)

As stated earlier, Ruby and its associated ecosystem (e.g., Rails) simplified the early design and implementation of the platform but were a bottleneck in terms of performance. In addition, as an interpreted language many errors are only revealed at runtime, which has an impact on system reliability. Runtime issues of interpreted programs can be detected through tests with adequate coverage, but the Twitter application did not have great test coverage at that time and there was no guarantee that coverage would improve quickly.

Twitter's engineers started to discuss which language to adopt to replace Ruby and decided to invest in Scala. This was not an obvious choice at the time—Scala was in its infancy. However, it was done for several technical reasons:

- The use of immutable values and the functional programming paradigm make programs more robust and easier to scale.
- Scala programs are compiled into binary code and have faster execution speed.
- The syntax and semantics seemed close enough to Ruby that engineers would be able to understand and ramp up quickly.
- The JVM (which is the execution platform of Scala programs) is a very robust piece of software that had, at that time, already matured over more than fifteen years.
- The JVM could be fine-tuned and optimized for the Twitter workload.

Twitter invested massively in Scala at that time, and the company helped to shape the ecosystem of this new language (e.g., with the introduction of the asynchronous programming, the development of the RPC system finagle, or the finatra framework). This new language provided the foundation to write reliable and scalable services. Applications were no longer single-threaded but were now able to handle multiple requests simultaneously.

Increasing Test Coverage and Continuous Integration

Engineers started to increase test coverage and make use of a continuous integration (CI) mindset and CI tools. They also started to have end-to-end testing, especially useful for the front-end and to avoid the WSOD mentioned above.

There were immediate benefits to this. First, as the company was migrating from Ruby to Scala, more errors surfaced at compile-time and hence not at run-time. This meant that many potential errors were not even deployed.

The second benefit was to have better control of what deployed. By having frontend testing (e.g., to make sure that some pixels or graphic artifacts appeared on the screen), it was then possible to make sure that some basic functionality was working before deploying a new version.

Ultimately, Twitter increased its testing coverage, providing more confidence that a change did not introduce a regression, as we have discussed as a best practice for addressing testing debt in chapter 6.

Looking Back: Recommendations from the Experts

The Twitter engineer that we interviewed mentioned that it is obvious that every project will have some technical debt at some point. We were interested to learn his recommendations to practitioners: not only to handle such technical debt but also how to avoid it in the first place. Here are a few of his recommendations.

Reduce Complexity and Skip Features You Do Not Need

The initial Ruby application had some very complex code (such as the multiple levels of caching mentioned before). This made the maintenance of the application very challenging, and it also made it difficult to migrate from Ruby to Scala.

The recommendation is not only to avoid complex code as much as possible (as recommended in the book *A Philosophy of Software Design* by John Ousterhout) but also to skip features that you might not need. In fact, before committing your team to more complexity, it is wise to wonder if (1) the feature can be realized without incurring more complexity and (2) if the feature is even needed in the first place.

Document Everything and Document Early

One recommendation was to invest in documentation. In particular, smaller teams do not invest in documentation much, but it is very important to document how the system should be at design time and write down the initial technical decisions. That will help to later understand the system and fill the gaps between what is implemented and what was initially specified. Documentation is really important in an environment where engineers move teams and new engineers are added to the team regularly.

Observe and Monitor

When you are responsible for an application that is hard to operate and maintain, it is difficult to monitor and know exactly what the contributors to a specific problem are. Also, when an application does not behave as expected, it takes significant time to find out the component causing the issue. For that reason, a great practice is to set up an observability stack, get signals from your software, and monitor its behavior (as recommended in chapter 7). This will help engineers to pinpoint issues in the codebase.

Knowing the contributors is useful, but it is also very important to organize postmortems with teams involved in a particular incident. Postmortems help to communicate context on a specific issue; they are also important to get insights and recommendations from other engineers that either experienced such issues before or have different (and potentially more effective) ideas to fix a problem. Postmortems maintain an open communication between teams (reducing social debt) and build a culture that emphasizes collaborative problem-solving instead of finger-pointing.

Further Reading

During our discussion, Kevin Lingerfelt mentioned the following book on software complexity: John Ousterhout, *A Philosophy of Software Design* (Yaknyam Press, 2018).

If you want to learn more about Twitter technical ecosystem, the Twitter engineering blog is one of the best resources. It provides great insight on all the tools and framework built at the company. The engineering blog is available at: https://blog.twitter.com/engineering. The Kiji approach is documented in: https://blog.twitter.com/engineering/en_us/a/2011/building-a-faster-ruby-garbage-collector.html; Finagle in: https://blog.twitter.com/engineering/en_us/a/2011/finagle-a-protocol-agnostic-rpc-system.html; and Finatra in: https://twitter.github.io/finatra/.

7 Deployment Debt

Ninety-nine percent is in the delivery.
—Buddy Hackett

You can't manage what you can't measure.
—Peter Drucker

Deployment debt relates to all the shortcuts, errors, or mistakes that happen when deploying and operating a system. Some concepts and techniques detailed in this chapter are similar to those used in the DevOps/CI (Continuous Integration) community. No matter what label you give it, the most important aspect is to develop software and processes that help you deliver the highest quality product with the least effort in the most reliable manner.

Not all developers are familiar with best practices for deploying software at a large scale. *Software deployment* means that the compiled program is being delivered or made available to the final customer. We often call this the "product", or, if you deliver a web application, the "production environment." As you deliver a customer-facing product, it should be (ideally) reliable and bug-free. This is usually an aspirational, hard to achieve goal. But to get closer to that goal of reliability, you need to not only address debt in your development process but also in the deployment process (that is, the effectiveness of how you deliver the final product to the customer).

Deployment debt occurs when you deploy software in an ad hoc (often labor-intensive) manner, making it difficult to deploy rapidly and efficiently, with a minimum of human involvement. It is also creating chaos, since manual operations are a source of errors. Deployment debt is what can potentially turn a successful business into bankruptcy, which is what happened to

Knight Capital in 2012 when a manual error in a deployment caused a $400 million loss, forcing the company to go bankrupt. There, a deployment mistakenly flipped an old software toggle that significantly changed the company's automated trading algorithms (see Further Reading).

7.1 Identifying Deployment Debt

When you develop an application, you write code that is approved through a code review process and ultimately lands in a code repository (see chapter 5 for more details about this). At some point, the code is shipped as your product (whatever it is: an embedded system, an app on a phone, a website, or an application on a desktop or mainframe). We call this customer-facing environment the *production environment:* if this is an embedded system, this will be the physical device; if this is a web-service, this will be the API or web interface it will have. The important part is that this will be what your end-user experiences. Let us review some of the most common symptoms that indicate that you have some technical debt in your deployment processes.

7.1.1 Symptom 1: Bugs in Production

The most common sign of technical debt is when bugs are found quickly in the production environment by your customers. In this scenario, an engineer deploys a new version of the software and a few minutes later customers begin to notice bugs (by sending you emails or bug reports, and by calling your support desk). This is the most common story when there are immature deployment tools and processes in place.

One might argue that this is mostly due to a lack of testing (see chapter 6), but testing cannot catch all potential errors. At best, all you can do with testing is to simulate the operation of the system, check that the result produced by your software is correct, and then check for edge cases and exceptional conditions (e.g., the unavailability of a database). But as large-scale systems depend on many external services, it is impossible to exhaustively test all combinations of potential problems. For example, if your system depends on external services outside of your control (which is very common today as the world is gradually embracing architectures based on services in the cloud), you might have some issues when deploying a new version of your system, because you now make more requests and use more external resources, because the system is misconfigured with invalid credentials, and so forth.

7.1.2 Symptom 2: Bugs Are Found After a Few Weeks

Another common issue is to find some bugs a few weeks after deployment into a production environment. This is also very common (although slightly less common than finding bugs immediately after your release), and there are two broad kinds of issues that could generate such a scenario:

1. **Edge cases:** the particular feature or combination of features that contains the bug is in a part of the system that is almost never used, and therefore it might take days or weeks or even months before a user experiences it.
2. **Temporal issues:** a bug is related to some temporal constraint or resource usage. Suppose some of your code goes into an infinite loop consuming too much CPU, or a memory leak causes memory consumption to increase without bound. This crash may not happen until days after you deployed it (see box 7.1, below).

The fact is, finding such issues is very hard and some of these bugs cannot normally be found with testing in a cost-effective manner. In addition, as mentioned before, testing is never complete, and some parts of your system are almost certainly not being tested, thus leaving latent bugs. This is not just a matter of cost and time, although more extensive testing does take more time and cost more. Such problems will always occur. So your goal should be to build the necessary deployment infrastructure and processes to catch these issues earlier and, when you don't find them by testing, to mitigate them as quickly as possible.

7.1.3 Symptom 3: Impossible to Roll Back

Consider this scenario. It is Friday at 3 p.m. Developers rolled out a new version of the software at 10 a.m. and customers started to experience a bug that has not been reproducible since the deployment of the new version. The lead of the development team is going for a hiking trip over the weekend and is leaving in one hour. Nobody knows what is happening. Engineers are looking in server logs, but everything looks fine there. For the sake of your developers' sanity, you propose to roll back and redeploy the previous version, but nobody has the old binaries. An engineer tries to rebuild the previous binary from your code repository, making assumptions on the date the previous version was released. The software is released. Unfortunately, the database schema has changed with the new version and also needs to be rolled back. Some engineers try to do this and make a mistake while modifying the database. Everything is down.

Box 7.1

Once upon a time, I was working for a company that deployed a critical service at scale. This service was serving tens of thousands of customers and had to have high availability. Bug fixes and new features were being shipped periodically, and we were rushing to deploy new versions on a constant basis. On a Friday night, a developer started to notice some high memory usage on a new version that had just been rolled out.

After a few hours of investigation, it turned out that the new version had a memory leak, and that all hosts were about to run out of memory and would start swapping. Needless to say, fixing the bug and deploying a new memory-leak-free version became the highest priority topic on the team, and half of the team was assigned to rolling out the fix. Without this engineer discovering this issue, the service would have been very slow, with a significant impact on the customer experience. In that case, the real issue was the absence of metrics and alarms used to track issues across the fleet.

—JD

Does that scenario (or some variant of it) sound familiar? This is less common today as the industry has been increasingly investing in DevOps tools and practices (such as the use of continuous integration, blending dev and ops engineers, automated and better testing, etc.). But it still happens, and as a result failures are sure to happen. The question is not how to avoid such issues entirely but rather how to minimize their occurrences and their impacts. This highlights the importance of automated roll back.

7.1.4 What Constitutes Technical Debt in Your Deployment Process?

While the following list is not exhaustive, this is what we consider deployment technical debt:

- **Dependence on manual processes:** manual processes are inherently error prone. Deployment and rollback (to the previous or older versions) should be automated.
- **Flying blind:** there are no metrics (and observability system to observe them) that show how the system behaves and detects potential issues.
- **Inability to manage deployment velocity:** changes in the production environment need to be rolled out slowly, in an incremental fashion, as opposed to a big-bang strategy where all servers are updated at once.

- **Lack of staging changes:** changes need to be tested in environments that are as close as possible to the production (or operating) environment.
- **Reduced observability:** lack of metrics for observing the system behavior, understanding suboptimal solutions.

These items are explained in the next two sections.

7.2 Managing Deployment Debt

To manage deployment debt, we offer several strategies.

7.2.1 Separate Deployment Environments

The first thing to do is to create a separate environment that is used to predeploy your system. You may call this preproduction, staging, canary, or something else—we have seen various names in the industry for it. We will call it "*preprod*" for the sake of simplicity. This preprod environment should be as close as possible to your real production environment in terms of resources, library versions used, dependencies, and so on.

The idea is that you deploy your system in the *preprod* environment and let the new version of the system operate for some time to detect any potential unexpected bugs. That gives you confidence that your system is actually working correctly, does not have any issue with external dependencies, and does not have regressions with previous versions. This deployment should be your last line of defense before releasing a new version (your first one is extensive testing, the second one is good and effective code reviews, and the last one is safe and gradual deployments).

In reality, what that means is that you should in fact have *three* development environments:

1. **Development:** where developers test their systems and where everything might break.
2. **Preprod:** where everything is stable and should be as similar as possible to your production environment.
3. **Production:** which is customer facing, and should be solid as a rock.

Finally, your preprod system might not have as much traffic as your production environment. For that reason, you might want to also have periodic tests against your preprod environment to simulate the behavior and load of your production environment that will check that the system behaves as expected (e.g., X% of the requests are handled, availability is Y%, mean response time is Z ms., and so forth). In the case of developing an embedded system, this means to simulate usage in different environments (changing temperature, humidity, zone where connectivity is bad, etc.). In the case of real-time systems, it means to introduce noise and attempt to make the system fail (by pushing time boundaries, making the system respond beyond its worst-case execution time).

These separated environments can help to manage debt due to **lack of staging changes**.

7.2.2 Observability to the Rescue

The separate deployment environments are a first step toward a better deployment process. However, you need to track how your system actually behaves as it runs. In that regard, you need to collect and expose metrics about your system. What you really need is an observability infrastructure: a system able to ingest metrics from your deployed system and present them in a meaningful way.

You need to collect several types of metrics related to the health and status of your system (more on that in the next two sections) and represent them so that you can track normal and abnormal behavior. Metrics are as simple as your system uptime, available memory, or CPU usage. These metrics are retrieved from each host or instance running your system (which may number in the thousands) and are then exposed through an API or graphical interface. The key feature of an observability system is to show different percentiles of the metrics (50th, 99th, 99.99th) as well as the min, max, and average. That will show you the normal behavior as well as potential outliers (abnormal values and where they come from). For example, on the graph in figure 7.1, there is a sudden spike of CPU utilization. This information helps developers and engineers to understand the potential causes of an issue and where to look for more information (e.g., to look at the logs for that timestamp).

There are several observability platforms available today. If you deploy your system in the cloud, there is a high probability that your cloud provider has a built-in solution to ingest metrics and build custom dashboards. If you deploy your system on premises, you will need to invest resources to deploy your own observability platform (or use an external service that

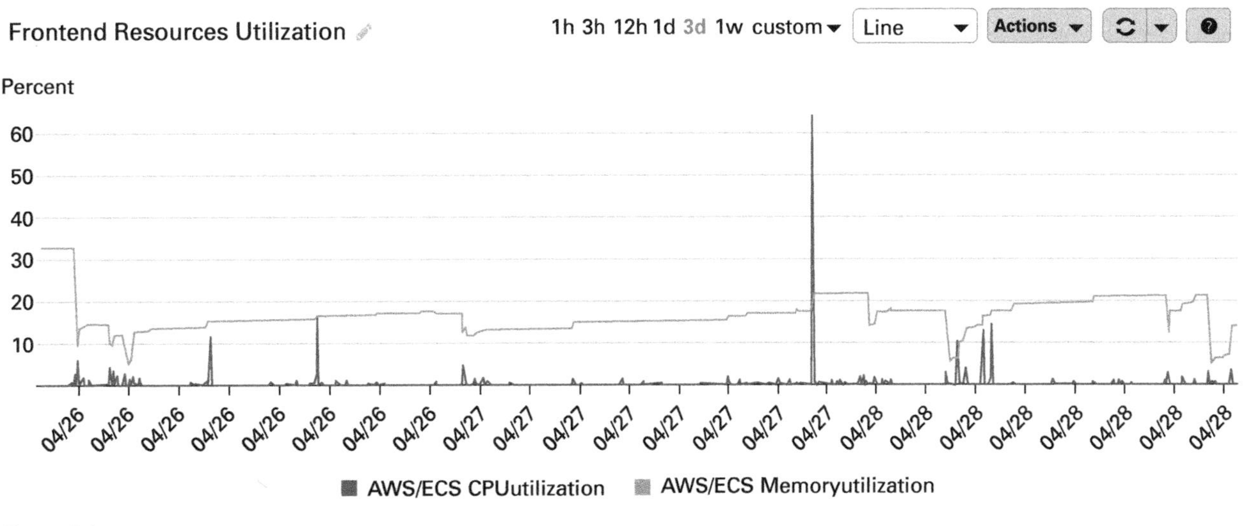

Figure 7.1
Example of metrics trends (CPU and Memory utilization) over a day.

ingests and surfaces your metrics—with the underlying problems in terms of privacy and security).

Once you have your observability infrastructure in place, you need to define what to track. There are two categories of metrics to track:

1. **Nonfunctional metrics:** CPU usage, memory, swap, uptime, number of processes, network traffic (in and out), latency.
2. **Functional metrics:** application-specific metrics (e.g., number of API calls, number of login attempts, number of requests of specific types)

Don't just use logs for observability An observability framework exposes clear, unambiguous metrics with a type (integer, float, etc.) and their semantics. This is better than simply relying on logs, and we strongly discourage the use of logs alone to observe and analyze a system. First, logs do not have structure and the log you are processing to observe your system may change, without notice, tomorrow. The second reason is that logs have extraneous information, and finding the right information requires processing of logs, which is often brittle (because log formats change frequently, and often with no advance warning), nonportable, and inefficient. Logs are best used for investigation into a bug or suspected attack and not for obtaining an operational overview of your system's health.

An automated observability approach helps reduce debt due to **flying blind** or **reduced observability.**

7.2.3 Track Nonfunctional Metrics . . .

Nonfunctional metrics are metrics that are unrelated to the specifics of your application functionality. They should be derived from the quality attributes identified in your requirements gathering process (chapter 3). Nonfunctional metrics focus on how your application is executing in its environment. What is important to ingest is everything that could reveal an issue in your system. Popular metrics to ingest are: CPU utilization, network traffic inbound and outbound, number of discards/errors on network interfaces, memory usage, and swap usage. If you are using any virtual machines to run your application (e.g., for a Java/Scala application), you should also monitor the behavior of this system: this is part of your execution environment! Thus you need to ingest the number of threads, the behavior of the garbage collector, and any other similar metrics that may reveal a lurking error.

The first step in tracking nonfunctional metrics is to choose which ones to track, which ones matter to you and to the success of your system. While we cannot give any universal recommendations, most stakeholders care about transaction throughput, response time, availability, and some aspects of usability and security. You need to figure out which of these metrics are correlated with success, however you define that. These metrics reveal two types of errors:

1. **Infrastructure related problems:** the capacity of your infrastructure is either too small (for example, your CPU utilization is high and you need to provision more resources) or your CPU usage is low and hence you are wasting money by overprovisioning resources.
2. **Potential bugs:** a new problem in your code impacts your infrastructure, such as a memory leak that causes a continuous increase in memory usage, or an infinite loop where the CPU gets completely consumed.

Thus you need to monitor these kinds of metrics. Now, you may ask: In what environment should I monitor them (development, preprod or production)? The answer is, of course: all of them! As you deploy your system, you need to keep an eye on all of these metrics and see how they evolve and change over time and across deployments and deployment variants. This will also show you the increase (or decrease) of resource consumption between different versions of your system and this data will help you to optimize your deployments.

7.2.4 . . . and Add Functional Ones

Functional metrics are less obvious and less general, because they depend on your application-specific concerns. Without knowing what system you are implementing, it is impossible to define what metrics to ingest. Nonetheless, there are some good practices to keep in mind.

First, have some metrics related to user behavior. For example, if your system is a web-based application, you should have a metric that tracks the number of login attempts, the number of attempts to recover a password (a high number might indicate an attack), and the number of users connected to your system. You should track anything that can reveal any potential abnormal activity.

Second, get metrics related to how your system interoperates with external systems. If you deploy a web-based application, monitor the latency and success rate of communications between each microservice. If you use an external service from a cloud provider, you should also track the success rate, latency, and return code for HTTP connections (e.g., number of 200 return codes—successful—and 5xx or 4xx return codes—failures). Interoperability metrics will also help you to make better diagnoses of potential issues. For example, if your system fails, the failure may not be a bug on your side but rather a failure of an external service. Of course, you should probably design your system to be resilient to this type of failure, or at least to fail gracefully.

Finally, you should think about how you might correlate functional and quality data. This will help you to highlight true outliers. For example, high CPU utilization might not be because of a bug but because of a sudden increase of traffic over an extended period of time due to a spike in legitimate usage.

7.3 Avoiding Deployment Debt

Finally, to avoid deployment-related debt, we offer several strategies.

7.3.1 Automated Gradual Deployments

Any manual activity is going to be error-prone at some point, and deployment is no exception. A big potential source of technical debt (in deployments and testing both) come from relying on error-prone human actions. Making a mistake when deploying a production service can have significant impact and might make your entire product unusable. For that reason, deployment should be done automatically. There is still the risk that you automate the mistake, but at least where that mistake came from will be easier to identify.

In addition to the automated aspect, deployment should also be gradual: do not deploy all physical instances of the same service at the same time. Instead, instances should be deployed gradually over time while metrics are monitored to detect a potential issue (e.g., regression, bug, etc.). For example, if you have a given service running on ten instances, deployment should start on one of them, then, after five minutes, if no metrics fall off, two other instances should be deployed with the new version, and in each five-minute period after that the deployment process could deploy twice the number as in the previous iteration until all instances are successfully upgraded. The benefit of making the deployment gradual is to avoid a full deployment of a version that will introduce any issue. A gradual deployment will limit the impact of a potential critical bug and make sure your system is still running (although potentially in a degraded mode, since not all instances are active).

The incremental deployment approach is also found in embedded software. For example, when Tesla (the car manufacturer) deploys a new revision of the car software, not all users receive the update simultaneously. The update is rolled out over multiple days or weeks.

If you are deploying your software in the cloud, all major cloud providers offer a service to automate code deployment, which is another reason to deploy to the cloud. If you manage your own infrastructure, you will then need to invest in maintaining external deployment solutions.

7.3.2 Integrate Deployment with Your Continuous Integration Pipeline

As mentioned earlier, one key principle is to keep separate environments for different stages (development, preprod and production). While the development environment should be the Wild West where developers deploy their own code, the preproduction environment should reflect the current status of the master branch and be automatically deployed.

When any service has a code change (from its codebase or any related library dependency), this service should be automatically updated in the preprod environment. The automated update will keep the preprod environment in a state that reflects the state of the current code and will help you to detect (and hence avoid) any integration issue as soon as possible.

The deployment from preprod to production should be automated but not automatically triggered. Ultimately, what triggers a deployment from preprod to production should be a *business* decision. Deployment to production should therefore be carefully considered and follow a well-defined process.

7.3.3 Define a Deployment Process

Even if the deployment process is automated, it does not mean that you should not pay attention to it or consciously define this process. Just the opposite: you should clearly document it (we cover documentation debt in chapter 8). All major technology companies have a special system to capture how their systems are deployed and to keep track of the state of the deployment. Some companies use a dedicated, custom-built system, whereas others rely on functionality from a ticket management system. No matter what strategy or tools you use, it is important to consciously and carefully define your deployment process.

Each component or service should have a deployment process. This process, and the documentation, must define at least the following aspects:

- What components are being deployed
- What action is being done to start the deployment
- What metrics the operator should monitor during the deployment to observe a regression issue and hence to potentially stop the deployment
- How to stop the deployment
- How to roll back

For each of these steps, the deployment process must explicitly define how to perform an action: either by clicking on an interface or by issuing some commands in a shell. This process must be documented so that the full knowledge related to the deployment is captured in a single document. Hence developers that deploy the system have all of the contextual information they need, whether they wrote the deployment process or not. This also helps any new developers on this project to have the same information as the original authors (at least for the deployment).

It is important to *follow* the deployment and record anything that happens (either successfully or not). A common way to document this is as a ticket in an issue tracker. The ticket should reference the deployment process and record each action being taken. Using such a system helps to track what has been done and when. This helps in any investigation during a postmortem.

In some cases, especially with embedded and safety-critical systems, it is not possible to remotely update the devices and instead, end-users must perform the update themselves. For example, some components in a plane are upgraded using a USB key that contains the update: the user plugs in the USB key and the system updates using the data on the USB key. In that particular case, it is critical to document the update and precisely give signals to the user: when the update is complete and successful, how to rollback, and so on. In addition, if the update is on a safety-critical system, it is also crucial to ensure that the update (file containing the new code) and the media containing the update (USB key in that case) itself are not compromised. This can be done by signing the update and using certified components.

7.3.4 Implement Kill Switches on New Features

Even when your code is fully tested, you might encounter issues after deploying new features. Issues may arise from the technical side (e.g., communications with external services, performance issues) or from the business side (e.g., users complain about a specific new feature).

For that reason, it is convenient to be able to integrate a *kill switch* (or *feature toggle*) for new features. The kill switch automatically disables a feature in your system at runtime, without having to make a new deployment. If the new feature has a problem and needs to be removed, a software-enabled trigger for the kill switch is sufficient and there is no need to roll back your software (with all the associated hazards and potential errors that a rollback can entail).

The kill switch should rely on very few other services. For example, if you implement your kill switch with a database, you might not be able to activate it if your system cannot connect to the database (for whatever reason). Similarly, the kill switch must be independent from the application itself (if the application has crashed, there is no way to activate the kill switch). One simple (and primitive) way to implement it is through files: the presence (or absence) of a file means that the kill switch is activated. While there are other ways to implement such a mechanism, the most important aspect is to keep it simple and decoupled from the main system and to document this information.

It is important to document the list of kill switches and periodically review if they are being used or not. An inactive or unused kill switch hides dead code that can potentially be activated by a bug or (more traditionally) a human mistake. For example, presumed dead code was one of the major causes behind the Knight Capital fiasco. It is therefore important to periodically review which ones are useful, remove the ones that are no longer needed, and make sure the code they are activating is removed (see the discussion about dead code in chapter 5).

For embedded systems, it can be difficult to implement such a mechanism, and safety and security concerns must be considered, especially if the switch is activated remotely (in that case, attackers could also activate the switch and harm the user). Local kill switches (e.g., a mechanical switch on the device, such as enabling/disabling a webcam on a laptop) are easier to understand and make it easy to reduce the attack surface. Needless to say, this should be discussed when designing the system and be approved by stakeholders.

7.4 Summary

This chapter explored technical debt beyond its typical definition and applies the concept of technical debt to deployment activities. Technical debt is not only code (although many deployment processes are code). In this chapter, we discussed technical debt in your deployment process and how it can affect your iteration cycles and time to deliver new features (value) to your customers.

This chapter explained the symptoms and consequences of technical debt in your deployment process and detailed the major steps to facilitate discovery of issues (observability) and deploy new revisions safely (use gradual deployment/rollback, use of kill switches). These techniques should help you speed up your iterations and deploy new revisions of your software faster.

At the end of the day, operating a system is like a car: it does not matter if you have the best product, what is important is to know how to operate it. In the case of the car, an operational (driving) error can wreck it. In the case of software, you can destroy a business in a small amount of time. The case of Knight Capital in 2012 is a great example: the issue would not have existed if there had been an automated deployment process, it would have been quickly detected if there had been metrics to observe what version was deployed on each server, or the bleeding would have stopped if there had been a kill switch. Because none of that existed, engineers did not know what to do, and the faulty system bankrupted the company. Investing in your deployment system is often seen as an investment that does not bring value (see chapter 11). Instead, it should be seen as an investment that will keep the company from sinking.

Further Reading

This chapter has focused primarily on identifying the technical debt aspects of deployment and is not a comprehensive guide to developing a full deployment process or defining your service level agreement (SLA) or service level objectives (SLO). If you are interested in these topics, we encourage you to read the *Site Reliability Engineering* book from Google. It is available online: https://landing.google.com/sre/books/. This is probably the best resource for system reliability engineers on operational aspects of a system.

This chapter references the software glitch that happened to Knight Capital in 2012. There is no authoritative document other than an SEC filing. Several blog posts examine it in detail. One of the best from a deployment point of view is Doug Seven, "Knightmare: A DevOps Cautionary Tale": https://web.archive.org/web/20200225104549/https://dougseven.com/2014/04/17/knightmare-a-devops-cautionary-tale/.

Good deployment practices are now being addressed by DevOps practices. If you are interested in this topic, we recommend the books *Effective DevOps: Building a Culture of Collaboration, Affinity, and Tooling at Scale* by Jennifer Davis and Ryn Daniels (O'Reilly, 2016) and *DevOps: A Software Architect's Perspective*, by Len Bass, Liming Zhu, and Ingo Weber (Addison-Wesley, 2015). To learn more about observability and more particularly how to achieve distributed systems observability (which is what you will need for any large-scale system), we recommend the book *Distributed Systems Observability* by Cindy Sridharan (O'Reilly, 2018).

If you want to learn more about good continuous development and continuous integration practices, we recommend the books *Continuous Delivery: Reliable Software Releases through Build, Test, and Deployment Automation* by Jez Humble and David Farley (Addison-Wesley, 2010); and *Release It!* by Michal Nygard (Pragmatic Bookshelf, 2018). If you are looking for a practical guide, there are books and resources dedicated to specific technologies such as *AWS Automation Cookbook* by Nikit Swaraj (Packt, 2017) or *Continuous Delivery in Java* by Daniel Bryant and Abraham Marín-Pérez (O'Reilly, 2018).

If you want to learn good practices on how to develop, deploy, and deliver embedded systems, we highly recommend Philip Koopman, *Better Embedded System Software*, (Drumnadrochit Education, 2010). His blog (https://betterembsw.blogspot.com/) is also a real gem for finding interesting information on famous (and infamous) issues in embedded software, such as the issues on the Boeing 787 or Airbus A350 mentioned in this chapter.

8 Documentation Debt

> Documentation is a love letter that you write to your future self.
>
> —Damian Conway

Documentation debt refers to shortcuts taken when documenting your software system, including outdated design specs and lack of code comments. Debt *in* the documentation is not the same as documents about the debt (documenting technical debt), which is the set of practices for capturing and describing technical debt.

Documentation is often seen as a thankless task that falls to some hapless newcomer on the team. But as anyone who has had to understand a nontrivial piece of software can attest, good documentation can save an enormous amount of time. The cardinal rule for writing documentation is that documentation should only be written when the *cost of writing docs* is less than the *value they might bring.*

Valuing the act of documentation requires some forecasting, of course. The principal forecasting exercise is to understand who might use the documentation and for what purpose. Most often, documentation debt happens when this forecasting exercise is wrong or not done.

In this chapter we discuss some ways to make your documentation more (cost) effective: figuring out what to document, making documentation a useful practice, and preventing common documentation mistakes.

8.1 Reasons to Document

Documentation is difficult to motivate in software development mostly because software developers aren't technical writers. For most of us, we work in computer science or related subjects because we love the technical aspects: perhaps the mathematics, perhaps the engineering of complex systems, or even the joys of finding bugs. Consequently, we have little to no training in writing for other people, and we frequently do not enjoy it. But technical writing is an endeavor of supreme importance, particularly when it comes to explaining the workings of complex systems. The other challenge faced by software teams is that the people best positioned to explain software decisions to others are often least interested or incentivized to do so.

Documentation debt exists any time you look at some code or a design and seek an explanation for why it is the way it is, and that explanation cannot be easily found. This means there is no obvious place where the explanation should live. The exact location varies, of course: it might be a comment above a difficult to understand algorithm, or a wiki page in the team collaboration server, or in a pull-request discussion. Documentation that requires us to track down multiple archival repositories to find the answer is not very helpful, as it increases the cost of using the documentation and the likelihood of making a mistake (because we may not be successful in tracking down all of the relevant information).

Documentation debt happens frequently: any time a decision is made that will be important for others to know (even oneself, a few weeks or months later: developers can forget the code a few days after they have written it) but is not captured for posterity, we create documentation debt. What we are not saying, however, is that pages and pages of diagrams capturing detailed behavior or documenting every class and method and variable need to be generated. Anything a competent developer should be able to easily recapture from the actual source code is unnecessary to document. Documentation debt happens when we choose, either for expediency or lack of interest, not to document something we know will be important later.

Documentation debt also happens when we create ineffective documentation. The classic software engineering paper "A Rational Design Process: How and Why To Fake It" (Parnas and Clements, 1986) lists seven key principles for effective documentation. Violating any of these principles creates documentation debt.

1. Write for the reader.
2. Don't repeat yourself.
3. Avoid ambiguity.

4. Use an organizing schema.
5. Record rationale.
6. Keep documents current but not too current.
7. Review documentation suitability.

Let's examine each of these in detail.

8.1.1 Write for the Reader

Good documentation works for a reader and makes the reader willing to read it. It is written in language the reader will understand. Is this document for end-users? Make sure every acronym is explained. Perhaps the readers will have graduate degrees in high-energy physics, and you do not need to explain electron spin. There are three major uses for documentation: analysis, construction, and education. We document so that we (or someone else) can, at a later time, analyze the system to understand it or to understand how it might support some proposed change. We document a design so that we can build a system that realizes the design (after appropriate analysis) in much the same way that blueprints for a building are the guides to construction. And we document to adequately and efficiently educate the newcomer to the project.

8.1.2 Don't Repeat Yourself

Documentation debt arises when the same thing is documented in multiple places. Aside from the obvious problem of wasting effort by documenting the same thing twice, repetition increases cognitive burdens on the reader, and it greatly increases the risk that the documentation will eventually become inconsistent. Like with duplicated code, there is no guarantee the writer will update each documentation duplicate. For example, the rationale for a design choice should only be captured in one place. That way everyone knows where it lives (and, remember, it might live with your source code).

8.1.3 Avoid Ambiguity

Ultimately we want to keep readers using the documentation. One reason is so they don't bother the people who originally wrote the documentation. Documentation can be a real force multiplier. But documentation that is ambiguous is worse than bad: now we have this potentially useful entry on the decision we care about but no clear answer to our question (should it be int32 or int64? Big-endian?). Specifications are notoriously ambiguous, and often it is only in the implementation that the ambiguity can be resolved. This ambiguity led to Postel's law: be conservative in what you do, be liberal in what you accept from others. In other words, strive to be unambiguous in what you write but liberal in your interpretation of other work.

8.1.4 Organize

Again, we want our readers to find the information they need. Debt is introduced when documentation is scattered and hard to find. Using a standard organizational structure for documentation allows readers to quickly know where to look. The most used places today are Wikis, collections of text files (that can be used offline and rendered online using a format such as Markdown, thus achieving the best of both worlds), or directly in the source code. We recommend putting documentation in the source code when appropriate (e.g., to explain coding decisions, such as why a particular algorithm was chosen) and into a set of text files to document other aspects (design rationale, collaboration process, and coding guidelines).

8.1.5 Record Rationale

Reading code can get one a long way to understanding how software behaves. One thing code cannot help you understand, however, is how or why a decision was reached. Why did they choose this particular algorithm or data storage approach? Debt is introduced here because the reader, who may be coming along months or years after the first implementation, will either have to re-reason the problem or may entirely miss some of the context for the choice. For example, the Square Kilometer Array telescope (SKA[1]) is designing the world's largest radio telescope, scheduled for science operations in 2023. However, the design decisions for critical parts of the operating software were being made in 2018/2019. Clearly, software teams coming to the telescope five years after the initial decisions were made will lack most, if not all, context behind the decisions. The SKA documentation approach will have to carefully capture the reasoning (a similar case study on a related telescope project, ALMA, is described in Case Study C). Capturing the reasoning now doesn't mean those later teams cannot change the decision: just that documentation debt exists if they can't even understand why the original decision was made.

8.1.6 Keep It Current and Consistent

One of the biggest sources of documentation debt exists when documentation is not current or consistent with the current state of the system (most often, the source code, but it could also mean new hardware or process changes). This destroys confidence in the documentation ("they said it used the Wronsky process, but we stopped using that ten months ago!"). At the same time, documentation that changes with every build also destroys confidence, because readers lose faith that the documentation is a fixed-point that can be relied upon. Documentation that is too current is probably either capturing things that don't need to be captured, like low-level coding decisions, or autogenerated on commit, so every tiny commit diff gets redocumented. Keep the documentation up to date on a release cycle basis.

8.1.7 Review

Documentation debt occurs when our documentation fails to meet reader needs. We need to not only think about who will be reading the documentation but also whether the documentation will meet those needs. That means figuring out whether the documentation itself does what we hope. Think of this like an acceptance test for documentation. We can think about a set of readers and then examine whether the current state of the docs would help those people. In chapter 5, rule 3 of efficient code reviews, "define the commit message format," also talks about the need to include documentation where reviewers know what to expect.

8.2 Identifying Documentation Debt

8.2.1 Estimate Needs

In many projects, particularly new or small projects, it is easy to write a lot of code before the need for documentation is apparent. In the furious pace of a startup or exploratory project, for example, motivating documentation is very difficult. After all, the company may fail or the project might get canceled next month! We return to the documentation value proposition: will documentation be worth the effort? We suggest the approach to documentation shown in table 8.1.

8.2.2 Identify Documentation Problems

Once you've located your likely documentation context (startup, corporate IT, regulated environment), there are a few obvious signs that the existing

Table 8.1
Project context and documentation approaches

	Project Context			
Documentation Approach	Exploratory / Startup	Late-Phase Startup	Corporate IT	Regulated Software
Documentation Priority	Low	Key to exit	Key to customer acceptance	Mandatory for regulatory approval
Key Design Choices	Lightweight capture—whiteboards, code comments	Textual capture, using design decision records stored in version control	Formally captured, using documentation management system	Formal capture, following industry standards such as ISO 42010
Code Standards	None/external	Language-specific using syntax like docstrings in Python	Commit checks and corporate standards	Contractors follow contractual standards, docs are part of deliverables
What to Document	CEO/CTO decide	Team leads/CTO decide	Documentation review boards	Everything

approach to documentation is not working effectively and that documentation debt is occurring.

Symptom 1: Lead developers answering questions The first sign that documentation is needed is often that experienced developers find themselves answering the same question about the system again and again. New team members come to their office because they don't know where to find the information in the wiki, issue tracker, or equivalent. A good rule of thumb is that the second time you tell someone the same information, it is time to write it down.

Symptom 2: You curse yourself The next sign documentation is needed is when you sit down to reread code from your own work, maybe code you wrote a few weeks before, and you start to realize you have difficulty understanding what you were thinking at the time. We very quickly forget the reasoning behind code we wrote just days earlier.

Symptom 3: Unnecessary documentation Documentation debt exists when the documentation that does exist is actively harmful. By "harmful," we mean developers have it inflicted on them, killing their productivity and failing to achieve whatever goal might have been set. Symptoms include mandatory documentation that lives in separate file systems from the source code; lengthy and context-free slideware everyone has to sit through; voluminous read-only PDFs that capture, in excruciating detail, elements of the system that are obvious from reading the code. A good check on unnecessary documentation is to first ask the developers what parts of the documentation they never use. In our experience, developers are quite frank about these issues! Another check is to use file-access logs to identify rarely visited directories. There are also emerging approaches that use natural language processing and machine learning to identify unhelpful code comments, such as a comment "//return the eigenvalues of the matrix" for a method called "eigenvalues(Matrix m)."

Symptom 4: Documentation aging Much like aging software, documentation gets outdated. If you read a design document or source code comment that causes you to realize the documentation is wrong, it is time to (a) replace and correct that documentation (or delete it entirely), and (b) devise some more rational process for ensuring the documentation is up to date. Deleting incorrect documentation, or marking it incorrect, is preferable to leaving it wrong. What we know to be wrong can change rapidly. What is wrong

Figure 8.1

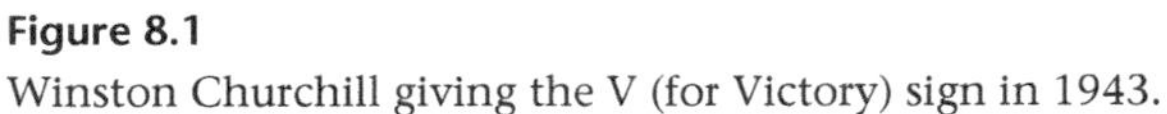
Winston Churchill giving the V (for Victory) sign in 1943.

might change when the person who created the documentation leaves or when the context for the original approach is lost. Thus it is safer to have no documentation than inconsistent or wrong documentation.

The objective is to make sure to avoid ambiguity. An analogy is the V sign, the meaning of which changes drastically based on context. Is it a sign of victory (as shown in figure 8.1) or a sign of peace? Depending on the orientation of the palm, it can also be an offensive sign. Words, pictures, and information depend on context, and you want to make sure your documentation conveys the same meaning through the years.

8.2.3 What to Document

Once we have a model of whether we need to document the system, the next question is what to capture. In table 8.1 we listed some scenarios that lay out

Table 8.2
Approaches to documentation

Artifact	Ease of Documenting	Speed of Change	Importance of Documentation
Code	Easy	Constant	Moderate
Tests	Moderate	Fast	Low
Design	Hard	Moderate	High
UI	Easy	Fast	Low

what is being documented based on what type of organization you work in. For instance, if you work on medical devices, your company almost certainly needs to adhere to US Food and Drug Administration (FDA) regulations on software, the FDA General Principles of Software Validation. These regulations usually contain specific instructions on what to document and the format that the documents should take. Contractual software development often specifies documentation standards as well as part of the deliverables. Failing to produce the required documentation will clearly cause business problems.

It is more difficult to know what to capture in other environments. Again, we return to the value maxim: if the documentation will not provide (estimated) benefits, then it is not worth capturing. There is no need to take on documentation debt if you can identify no need for the documentation in the first place. Table 8.2 lists one possible prioritization scheme. Most projects implicitly use the following documentation maturity model:

1. New project, no disciplined development, no documentation
2. Coding standards introduced; code comments required at merge
3. New design proposals required to provide design documentation and justification
4. Documentation systems introduced (wiki, documentation management, online API documentation)

In your case, you might choose to get started with those artifacts that are easy to capture but balance that against the importance (estimated) of that documentation.

8.2.4 Why Documentation Debt Is a Problem

Documentation is too often a second- (or third-) class citizen of software development. Managers let developers write the documentation once the code is written—or worse, when the code is shipped and developers are already working on another project and have lost the context of the project! Most developers or managers do not realize the importance of documentation for different artifacts in the software lifecycle and how important it is to your product visibility, both internally and externally.

From an internal standpoint, good documentation reduces time to onboard and learn the software. New hires (or people that were newly assigned to a project) will be productive quicker, and you will deliver new products and features sooner than you expect (or just on time). It also helps communication across the team: an undocumented API means digging into the code and spending hours attempting to understand how it actually works. This is even more important when working with distributed teams, scattered across time zones, or outsourced components.

From an external standpoint, appropriate, up-to-date documentation should be the norm. Failures in documentation for external users are interest payments on the technical debt introduced by short-cutting the required work. There is nothing worse than incomplete documentation (for example, it is common to have an API where only success cases are documented, and you need to figure out failure modes and failure behaviors). When bad documentation is shipped with the product, the user will be frustrated (at best) or just not use the product at all (at worst). There are numerous very powerful products that struggled because users did not understand them, so do not ship one of these.

For example, in 2011 Apple introduced iCloud Core Data, an object-level cloud storage and syncing API for apps on iOS and OSX. It was intended to compete with offerings from Google, Dropbox, and others as a way to sync content and settings across different devices. It soon became clear to external developers that on top of many bugs, the API was very poorly documented and was full of documentation debt. After many years of complaints, and doubtless lost revenue, Apple deprecated the entire API for iOS 10, replacing it with CloudKit.

Amazing documentation, particularly for external API users, has many facets. The best examples provide quickstarts, sample code, extensive documentation for each function, well-defined data types, and glossaries. There are numerous examples online (see Further Reading).

Good documentation can have a great return on investment: all it takes is one developer to write the documentation, and it will impact several developers/managers (internally) or users (externally).

8.3 Managing Documentation Debt

Having chosen what to document, and what documentation is missing, the next phase is to manage the documentation process. Managing documentation debt is about making the creation of documentation easier and keeping those docs in a consistent place.

8.3.1 Write Documentation Along with the Code

For a long time, with one significant exception, documentation and code have been seen as separate artifacts in software systems. This largely aligned with the waterfall model of development, with its insistence on separate, gated steps in the process. A lineage of software development from defense, government, and other procurement-driven industries further enforced a separate documentation phase, at times written by entirely separate teams or contractors. This led to the inevitable problems the Agile Manifesto railed against: a focus on large, unwieldy, and monolithic documentation instead of working software; documentation that at any rate, no one read in its entirety, except an acquisitions officer signing off on deliverables. Creating unneeded documentation is technical debt. See Further Reading for an example.

Over the years, languages and tools started to address the problems with documentation. The root of these changes likely starts in the CERN hypertext project that led to the World Wide Web. HTML and later wikis showed that a lightweight, text-dominant format could be rapidly disseminated. Indeed, the original purpose of Tim Berners-Lee's project was to make knowledge-sharing at CERN easier!

A related avenue of work was the concept of Literate Programming from Donald Knuth, in 1984. His vision was that code and documentation ought to be intertwined, so that changing one or the other was done in the same place. His tool for this, WEB, defined a language syntax for weaving documentation into the code. The code snippet below shows a simple example describing Knuth's code for binary trees.

```
\datethis
@* Introduction.
A binary tree is a {\it normal form\/} if it is the
representation of
some integer as described above. It isn't hard to
prove that this condition
```

```
holds if and only if each node $x$ that has a right
child $x_r$ satisfies
the condition $v(xl)>v(x_{rl})$.
@s node int

@c
#include <stdio.h>
@#
@<Type definitions@>@;
@<Global variables@>@;
@<Basic subroutines@>@;
@<Subroutines@>@;
@#
main()
{
register int k;
register node *p;
@<Initialize the data structures@>;
while(1) @<Prompt the user for a command and execute
it@>@;
}
```

This snippet demonstrates the principle Knuth passionately believes in: that separating the explanation for the program from the code is a mistake. Similar mechanisms, although not as powerful, can be found in most modern language infrastructures:

- JavaDoc annotations to support document generation in Java code
- Godoc will take any comments preceding a block and extract them as documentation
- Python docstrings define a location and syntax for adding inline method documentation

These tools, while separate from the language itself, are tightly integrated into the language infrastructure. One wouldn't be able to use Python-style code comments with Lisp, for example.

To solve the problem introduced by polyglot programming (the practice of using multiple languages in a single system), documentation support tools are important. One of the oldest and most popular is Doxygen, first released in 1997, which extracts annotations from the code and generates documentation in whatever format you need. A number of similar tools have sprung up to support documentation, including Sphinx and ReadtheDocs. All tools are similar in their use of simple, plain-text formats such as Markdown, RestructuredText, or Pandoc.

Plain-text has two benefits. One, it makes extracting the documentation into other formats very simple, which avoids vendor lock-in that formats such as PDF offer. Most importantly, these formats, since they are simply text, can be stored and versioned in the same repository as the code they document. While not inline—like Knuth suggested long ago—keeping the documentation closer to the code makes it more likely that the documentation is actually updated:

1. The documentation is under review at the same time as the code so that documentation is part of the review and development process.
   a. No documentation written with the code = no approval.
   b. Code approved means documentation approved.
2. It is easier to review the documentation (as it is embedded with the code).
3. The documentation can be versioned with the code (therefore, you can keep track of the history of the documentation and regenerate old versions if needed).

We explored this in detail in chapter 5 when discussing effective code review, but it is important to consider documentation as a first-class citizen, at the same level as code and therefore, requiring the same approval and validation processes.

The remaining challenge is where documentation that goes beyond a single code file should live. That is, clearly the documentation for the User class can live either in the source code (filename User.py), or in a separate documentation file (for example User.md). But where should the documentation for the entire business logic tier live?

One approach is that this overarching documentation should live in a higher level folder, within the same project, such as the following:

```
/src /doc /bin /test
```

Within the /doc folder, create a directory, /arch that holds architecture decisions (what we called design rules in chapter 4), and project-wide quality attribute models. This is the approach Michael Nygard uses for his architecture decision records.

Another model is to create a docs repository and keep the less frequently changing design documentation in this separate repository. This makes it harder to keep these current with the changes, as we outlined above, but does help to separate commits to code from changes in documentation. This might make your source tree cleaner, particularly if there is a separate documentation team. This is the approach used by TensorFlow on GitHub, for example.

8.3.2 What about Diagrams?

In our experience, for teams that do use diagrams and notations such as the Unified Modeling Language (UML) it is rare that these diagrams are updated. Often they are used in a slide deck for the initial design review, then ignored thereafter. However, even teams that do not do significant modeling with diagrams will sketch designs on whiteboards and scrap paper. These diagrams may have significant value. For example, they serve as a historical record. To ensure the possible technical debt associated with UML and design diagrams is minimized, we recommend keeping the storage format as simple as possible. Storing the diagram in a proprietary format (e.g., Microsoft's Powerpoint, UML tool formats) is a risk. We suggest using popular and standardized formats like image files (PNG, JPG) or PDF. If you do use a more specialized diagramming tool, we recommend picking a notation that many people will understand and ensuring that you store it in a standard format, such as XMI.[2] Storing something, even a back-of-the-envelope picture that you scratched out and photographed, is better than storing nothing. These diagrams often serve to bring people together, to focus discussions, and to aid in exploring alternatives.

8.3.3 Write Code, Tests, and Documentation All Together

A final approach is to keep tests around as low-order documentation. Very often code, documentation, and tests are out of sync: the documentation is outdated so that some newer functions are not there, or some tests might not work (if you do not have any continuous integration system in place, from chapter 7). This is technical debt and you will need to pay it off later (by updating the tests and/or the documentation). The longer you wait the bigger the debt. This is because, at best, you will try to adapt code and documentation related to a change that you did a long time ago and will require some time to remember what this change was about. At worst, the original author is long gone and you need to do some archaeology.

Writing tests (as we described in chapter 6) can help with documentation, particularly if the tests are written using frameworks like Behavior Driven Development (BDD), which focuses on capturing business-oriented acceptance tests. BDD uses the "given-when-then" framework to record *why* a particular test is being run.

Some languages provide the capability to embed documentation and tests in the code and to validate them with the compiler or build system. That ensures that your documentation and tests stay synchronized with any code change. The *rust* language provides this capability, as shown below: you can write documentation that contains some tests and show how to use the function. In the following example, we define a function, fib; show how to use it; and verify the correct values from the second Fibonacci number to the fifth.

````
// Compute the n-th number of the fibonacci suite
//
// This is a very naive implementation to find
// the n-th number of the fibonacci suite.
//
// ```
// let n: [i32; 5]=[2,3,4,5,6];
// let r: [i32; 5]=[2,3,5,8,13];
//
// for number in 0 .. 5 {
//    let result=fib::fib(n[number]);
//    assert_eq!(result, r[number]);
// }
//
// ```
pub fn fib(n : i32) -> i32 {
   return match n {
   0 => 1,
   1 => 1,
   _ => fib (n-1)+fib(n-2)
   }
}
````

The rust build-system provides the ability to auto-execute these tests (among other unit tests), as shown below, making sure that any code that is shipped with the documentation passes. Such functionality is available in different languages and tools. Use it as much as you can.

```
$ cargo test
Compiling fib v0.1.0 (file:///home/julien/
technical-debt-book/examples)
  Finished dev [unoptimized+debuginfo] target(s) in 0.50
  secs
  Running target/debug/deps/fib-333ae8c32de5f528
running 0 tests
test result: ok. 0 passed; 0 failed; 0 ignored; 0 measured;
0 filtered out
Doc-tests fib
running 1 test
test src/lib.rs—fib (line 6)...ok
test result: ok. 1 passed; 0 failed; 0 ignored; 0 mea-
sured; 0 filtered out
```

Like storing documentation in the same folder as the source file, the challenge with using tests as documentation is that some of the documentation does not have a natural mapping to any particular test and needs to be stored in another location. Another challenge is that the tests either capture obvious facts or repeat the same fact over and over again.

8.4 Avoiding Documentation Debt

Remember that our first principle for avoiding documentation debt is to ensure we only document what we think will benefit us *more* than the effort to document will cost. So if we can avoid writing documentation that no one wants, we immediately ensure our return on effort will increase. How can we determine ahead of time which documentation is useless or unnecessary?

Our opening section in this chapter mentioned several ways, following the seven principles of effective documentation. This includes figuring out who your reader is, and ensuring some consistency and lack of repetition, and writing things down that are frequently asked about. But we want to suggest two other possible approaches to understanding the effectiveness of your documentation.

8.4.1 Traceability

The first approach is based on the notion of traceability that many software development and continuous integration (CI) tools now support—and which we discussed in chapter 3. Traceability traditionally referred to the notion that we can identify a chain from requirement down to source code commit. Often this was so that in safety critical systems, like medical devices, there was a guarantee that someone had thought of a particular requirement. Most of the time this traceability is implemented manually, in traceability matrices listing requirements and software modules.

In modern software development tools, this traceability to some extent comes for free. Issue trackers such as Jira or GitHub allow for tagging of commits with issue ids automatically (e.g., "closes #23"). This traceability is an excellent way of understanding the connection between features you might be developing (either captured as user stories or feature requests) and the code that implements that feature. Effective documentation should enable this end-to-end reasoning, particularly in a continuous, rapid feedback world. A lack of traceability, or even occasionally omitting traceability, reduces your ability to understand what features are in production and which ones are successful. This is even more important when measuring results such as performance metrics.

8.4.2 Check Quality of the Documentation as Part of the Release Process

As your software evolves, you still need to keep improving your documentation and make sure it is up to date. We have seen that any code change that impacts documentation should come with documentation changes as well. However, this means that only documentation related to code changes is updated and does not give any insight about the overall documentation quality. As for everything, you can only improve what you measure, and you need to put in place metrics on your documentation. Let's see some of them.

The first one is to get a documentation linter that looks for any language (natural or programming) or cosmetic issues. Documentation generators include modes that report errors in the documentation such as dead links or syntax issues. If your linter does not include vocabulary/grammatical verification, use free dictionary tools like GNU Aspell. That will give metrics on the number of issues related to the documentation itself, and your objective is to decrease this number over the software lifecycle. Any increase should be remediated immediately and developer spare-time might be used to address existing errors.

The second metric to watch is the age of the documentation: using the configuration management system, you can track the age of each section and then check if old sections are up to date. Maintaining a metric such as the minimum, maximum, and ninetieth percentile of the documentation age is a good idea and will give insight on pages to update.

The third metric to watch is the documentation coverage: how much of your software is actually documented. This is hard to measure for user documentation but can be done easily if you develop an API. This metric should always increase in the software lifecycle and any regression incurs technical debt. However, documentation coverage is a quantitative metric that does not guarantee documentation quality. Quality is ensured by reviews from the developers; that is the reason why documentation changes must follow the same process as code change. They must be reviewed and approved to be shipped. Recall our final principle: review your documentation to ensure it meets your needs.

8.5 Documenting Design and Technical Debt

Another aspect of software documentation is how to document technical debt. Documenting technical debt is different than technical debt in documentation, which is what the preceding parts of this chapter looked at. Documenting debt is important to make the technical debt transparent, to prepare for repayment, and to ensure it is being undertaken strategically. We have the following suggestions for best practices in capturing technical debt.

1. First, and perhaps most obvious, is that you should plan to explicitly document technical debt. This might be as simple as a new label in your issue tracker. Without being explicit, it is impossible to plan for removal.
2. Capture the debt as it is incurred. Deliberate technical debt is a legitimate development strategy, but it is important that these shortcuts be captured for later. So-called self-admitted technical debt, in the form of code annotations such as FIXME or HACK are a poor second cousin. Extract those comments into proper issues. See box 8.1 for an example.
3. Sprint or iteration planning should leverage and allocate part of the sprint to dealing with debt, refactoring, and cleaning up the codebase.

Box 8.1
Voice of the Practitioner: Marco Bartolini

> One of the difficulties we face as a project [building a large telescope] is represented by the quality of some highly used scientific libraries. Some of these software products are the state of the art in terms of functionality, but they are often the outcome of layers and layers of research work, often resulting in poor maintainability and testability of these systems. But science libraries of new generation, such as astropy, really shine under the aspect of software quality as well, so our life is becoming easier. When we accept to use something knowing its limitations, we usually create one or more features in our backlog representing the work needed to refactor or rewrite the product according to our quality standards. These features follow the natural lifecycle of estimation and prioritization as any other feature of the system, but we have them in a dedicated bucket (an epic in our system), so that we always know what's the total amount of debt and we can decide to force some allocation if we see it is diverging.
>
> —MB[3]

How you do this varies. In some companies, an entire sprint (post-major release) is dedicated to clean-up. For others, debt is something you deal with as you read and revisit code (similar to a chef keeping a clean counter and kitchen). Some companies allow time each week explicitly for debt remediation.

4. As part of your design specifications, architecture decision records, or other documentation of the design, add an explicit technical debt section. This might be part of the rationale or completely separate. This accomplishes two things: One, it acknowledges explicitly, for senior stakeholders, that some amount of technical debt is expected. And secondly, it supports reasoning by later readers, who may suspect that debt exists but not know if it was deliberately incurred or why.
5. Finally, as part of your architecture documentation, dedicate a section to a mapping from modules of the software to existing technical debt items. Many tools exist that can show technical debt. However, the tool's definition and yours likely differ in significant ways.

8.6 Summary

Documentation debt is a significant problem, although it might not be what you think of when we talk about technical debt. Anyone who has had to work with software where documentation shortcuts were taken, such as poor, out of date, or nonexistent documentation, can attest to this. Documentation should only be produced or maintained if the effort of doing so will be less than the value that documentation can be expected to produce. It is critical to understand what you should be documenting, particularly by understanding for whom you are documenting. Keeping documentation modular and close to the code it is addressing is a key technique for reducing documentation debt.

Notes

1. http://skatelescope.org.

2. https://www.omg.org/spec/XMI.

3. The full text of this and other Voice of the Practitioner sections can be found in the Appendix under the relevant name.

Further Reading

The book *Documenting Software Architectures*, 2nd ed. by Paul Clements, Felix Bachmann, Len Bass, et al. (Addison Wesley, 2010) is an excellent overview of the challenges and approaches to architecture documentation. They borrow the seven principles for documentation from an excellent paper by David Parnas and Paul Clements, "A Rational Design Process: How and Why to Fake It," *IEEE Transactions on Software Engineering*, SE-12, no. 2 (1986): 251–257. That paper is well worth reading: it explains how even though your process may not actually be rational, retrofitting a veneer of rationality, such as with proper documentation, nonetheless pays off.

Michael Nygard, who wrote the excellent book *Release It* (Pragmatic Bookshelf, 2017), first introduced the notion of Architecture Decision Records in his blog post: http://thinkrelevance.com/blog/2011/11/15/documenting-architecture-decisions. The ReadTheDocs community is a good source of new documentation ideas, particularly from a technical writing standpoint. They host an annual conference called Write the Docs.

An example of documentation linting is nitpick for sphinx: http://www.sphinx-doc.org/en/stable/config.html#confval-nitpicky.

There are a variety of types of regulatory documentation requirements. For example, the FAA and others adhere to DO-178B/C, but the FDA is the example we use here: https://www.fda.gov/regulatory-information/search-fda-guidance-documents/general-principles-software-validation. Compliance with the standard is necessary to achieve certification (e.g., to fly a plane or sell a medical device) but typically results in long and unwieldy documents.

About the issues mentioned on iCloud and Core Data, Michael Tsai captured developer problems with Core Data in this post, and follow-ups: https://web.archive.org/web/20160318012147/https://mjtsai.com/blog/2013/03/30/icloud-and-core-data/.

The most easily available large-scale documentation effort might be the A7 examples, described in the system requirements specification here: https://apps.dtic.mil/dtic/tr/fulltext/u2/a255746.pdf. There are other examples, using for instance the outdated MIL-STD 498, such as: http://web.mit.edu/16.35/www/project/297749RevF.doc.

Martin Robillard and Robert DeLine, in "A Field Study of API Learning Obstacles," *Empirical Software Engineering* 16, no. 6 (December 2011): 703–732, report on a comprehensive study the authors conducted at Microsoft on API documentation and learning. They discuss common problems developers had with using APIs.

If you are looking for good, well-written, and useful documentation, the Amazon Web Services (AWS) documentation is a great example. Their boto3 documentation (Python library to interact with cloud services) is an example of good documentation, with each function and class being documented with examples, and an extensive list of success and failure cases. Another case of good documentation is the Stripe API that is well known to be developer-friendly and also easy to use.

Finally, literate programming continues with a small but active community, hosted at http://www.literateprogramming.com/index.html. Recent developments with computational notebooks, such as Jupyter, seem to be reviving the idea of mixing code and documentation, for data science purposes.

Case Study C: Scientific Software

Summary and Key Insights

Scientific software runs some of the largest, most complex systems in the world. From the Large Hadron Collider to the proposed Square Kilometer Array, scientific software processes terabytes of data, across thousands of kilometers, often in hard real-time contexts. And yet scientific software traditionally was developed by domain experts (physicists, chemists, astronomers). As a result, fielding, commissioning, and operating these systems have involved a number of technical debt workarounds. In this case study, we describe one such system, the Atacama Large Millimeter Array (ALMA), currently under active use in the Chilean desert. We look at how the ALMA project faced multiple sources of technical debt, including in requirements, code, design, and testing, and how they have during the years before and after the instrument went live dealt with this debt. In particular ALMA, and many such projects, face social debt challenges, as the projects are very long-lived, run across many national organizations, and see a shift in team structure between telescope development and telescope operation.

Background: Radio Astronomy Software

Radio astronomy software is the very definition of niche, but in fact software in these systems is at the cutting edge of software engineering practice. Many important developments have come about because of scientific data processing needs. For example, the World Wide Web was created at the CERN high-energy physics lab to address communication and collaboration problems.

In radio astronomy, large antennas are used to image space at the radio wave portion of the spectrum (1 mm wavelengths to 100 km wavelengths). As with any electromagnetic radiation, we can see new things using these wavelengths.

The first radio telescopes were large dishes—like the one in the James Bond film The Living Daylights (1987), which is the Arecibo Observatory in Puerto Rico, or the Very Large Array telescope dishes Jodie Foster's character uses in Contact (1997). As the science became more refined, more and more data was gathered. The dishes became wider, and new types of antennas—even virtual antennas made from several individual antennas—were commissioned. To deal with the larger volumes of data, scientists turned to software processing, for assistance. Nowadays, the largest and newest telescopes in the world run some of the most powerful supercomputers to process the terabytes of data they receive every day.

There are a few unique factors that make technical debt interesting here. One, these telescope projects are publicly funded and usually underresourced. The people doing the software engineering are as talented as any engineer, but there are few of them. Two, these telescopes are expected, given their large public cost, to exist for decades or more. Finally, the telescopes exist in a complex problem space and are usually tackling problems no one else has encountered. It is common for new designs to involve leading engineers at the largest Internet companies to brainstorm future designs for processing vast quantities of data.

Software Process for ALMA

The ALMA is a set of sixty-six radio dishes in the high Chilean desert, run by the ALMA Science Foundation (an international partnership of national science foundations). As with most large, billion dollar science projects, planning for the telescope began decades before science operations started in 2013. Construction began in 2004, and software development and design started in 1999. In telescopes, major phases can be thought of as four distinct time periods. First, there is *planning and design*, sometimes called preconstruction, which is a multiyear process of soliciting funding, contacting domain experts and people with past engineering expertise, and creating the set of conditions to allow for the next phase. *Construction* is broadly the phase where physical construction begins: roads, fiber, power, and eventually building and installing the (12 meter) dishes. *Commissioning* is the penultimate phase: this is about testing and initial deployment of the science instruments of the telescope. This means running tests, bringing one or two dishes online to test signal processing, getting the supercomputer cluster up, and of course, installing and deploying initial versions of the software. Finally, *operations* is the phase when the telescope is handed off to the operations team, and scientists can begin submitting requests to have the instrument conduct observations (e.g., to observe a particular portion of the sky at a particular time and wavelength).

This is a waterfall process, as with most cyberphysical systems. Decades of experience exist in systems engineering and planning for construction of large structures. What is different in the case of ALMA, and similar instruments, is the need to simultaneously build the software infrastructure capable of processing the observations. Particularly in radio-astronomy, software is needed to correlate the various dishes to create a single, powerful instrument. The software processes raw data (in the form of radio observations) and correlates those observations with those of the other instruments. This is an extremely data and computation-intense operation.

Software at ALMA has three main roles. The first is to observe and process incoming data and translate it into usable, postprocessed brightness data for scientists. This part requires strict real-time processing guarantees. The second role is to manage the observatory: point the telescopes, run the supercomputers, manage security and login at the site, and monitor key data points and error conditions. Finally, software is used to manage the scientific process itself. Most users of ALMA do not travel to the physical site but submit scientific requests to a central committee who evaluate the value of each request and then schedule the telescope in the most efficient manner. As of the writing of this book the entirety of the software code base is over 2 million lines of code.

Technical Debt at ALMA

Since planning for the telescope software began in 1999, and actual operations began in 2013, that leaves a lengthy gap between software design, development, and deployment. A fourteen-year gap is an unbelievably slow rate of deployment by today's standards. As a result, it is fairly easy to come up with a design in 1999 that is totally outdated in 2013. For example, volatile memory in 1999 cost $1.25/megabyte, and $0.0061/megabyte in 2013—a factor of 200. Such drastic reductions have significant system design implications. Since the design must accommodate the potential hardware available when construction begins, often the telescopes are designed by assuming that hardware capabilities will continue to increase—say, using Moore's law of exponential growth of processing speeds. As a result, when it comes to testing, it often cannot be done on the actual hardware but instead on models of the hardware that are predicted to exist.

The software engineers in charge are not ignorant of the need to deliver incrementally and gain feedback, of course. As a result, early prototyping before commissioning was heavily used. Another benefit is that for some functions, such as scheduling, lessons and occasionally code from other telescopes can be reused. Many of the personnel are former or current graduate students who have worked with earlier telescope projects. As a result, there is excellent knowledge transfer from project to project.[1] This transfer happens not only as staff move but also as reports and papers get written as part of the scientific process that underpins these projects.

We identified the following categories of technical debt in the ALMA project: social debt, requirements debt, code and design debt, testing and deployment debt. While more might exist, such as documentation debt, these were the ones that were well-grounded in reports and public evidence. We detail each one below.

Social Debt

Since the telescope is a large, multinational collaboration, there are many cultural and logistical issues making collaboration challenging. For example, it is hard to manage video conferences given the many time zones involved. Mitigating against this is the way in which the teams are highly motivated to get the instrument operational for science reasons, and the community is reasonably tight-knit, meeting at annual conferences. The software is built by a large number of teams, each responsible for a subset of the overall functional architecture. This worked well in the initial stages, but as cross-component issues became more important (such as quality attributes like performance), there was need to collaborate across subsystem teams.

To resolve this, ALMA introduced function-based teams, which managed requirements that cut across subsystems.

Requirements Debt

The requirements for these large projects are typically well-defined in advance, and the telescope, while novel, contains large numbers of well-known requirements from other projects. At ALMA these requirements were managed with a formal requirements tool and change tracking. The biggest source of requirements debt was scope creep, as once one requirement was implemented, end users could then begin asking for improvements elsewhere (e.g., making the software more efficient once it stopped crashing). ALMA requirements for new capabilities (e.g., capability upgrades) are now managed with a formal scientific change control board and roadmap process.

Code and Design Debt

ALMA is written in C (real-time), Java (control software), and Python (science interfaces). To test and prototype the design, a small test facility was used to support software spikes, architectural ideas that could be piloted on the smaller array. Once those lessons were learned, the actual design was fairly straightforward. Early on a decision had to be made (the last responsible moment had arrived to make the decision) that CORBA was the middleware of choice for connecting software components (late 1990s/early 2000s). In hindsight this was probably not a good choice, but it isn't clear what could have replaced it (at the time). Now that the telescope is active and conducting science, new designs for improved capability are driven by prototype and test reports—for example, how to improve data processing capability with existing hardware.

These technical decisions, often highly constraining technology, framework, and language choices, typically must be taken quite early, as they sit low in the technical stack. At the same time, since they are low in the abstraction levels of the system, they can be very difficult to replace. For a similar project, the Square Kilometre Array (SKA; see Interview: Marco Bartolini), a similar decision was required around messaging protocols. This is the operating system, if you will, for the distributed systems of the telescope. In this case, the SKA project chose Tango but only after months of prototyping and testing, albeit not the eventual scale at which Tango would operate. However, precursor telescopes using Tango provided another, larger-scale practical test.

In the construction phase, performing initial, proof of concept science observations is the critical goal. The telescope has been in development for many years, and the ability to do science with the instrument is tangible.

This pressure led to software standards slipping to make the deadline: code cloning, outdated libraries, and adding workarounds instead of fixing the real problem. While the debt is tracked in a backlog, the debt is not the immediate concern, and the commissioning team is aware that a subsequent team will be in place to manage it. The operations phase, which has a smaller software team, is now responsible for fixing this debt. They are, to this day, still struggling with the debt. However, the CASA software package, which manages data processing, is (now) a mature, open-source set of tools that other observatories are using. CASA follows a well-defined process for releases and development.

Testing Debt

Testing such long-lived systems is very challenging. While simulations can offer some evidence, without the real, high-data-rate system, conditions will always conceal some of the complexity. For example, in large data stores, some portion of the data gets corrupted due to cosmic rays flipping bits, so backups and error correction are vital. But this is difficult to test in simulations.

What worked for ALMA was to insist, in an engineering-focused project, that the software infrastructure be given real hardware (prototype dishes) to experiment with. However, this is not always easy to balance against the perennial demand by scientists—the ultimate stakeholders—to dedicate real telescope capability for science and not software testing, no matter how useful the end product will be.

Delivery Debt

A big challenge for ALMA was moving from isolated yet functional components, to an end-to-end pipeline that delivers useful science data products. In each component there are inevitable assumptions that certain inputs or outputs will exist, and these are mocked while development proceeds. As a result, the ALMA team created a separate integration testing facility, and during the handover to commissioning teams (the ones who are operating the system), they worked in parallel to debug the end-to-end issues that arose. Another approach the ALMA project used successfully to avoid debt was to standardize on a common set of infrastructure and software tools.

Technical debt was frequently created during commissioning. This is an active yet costly stage of the project: all the hardware is there and incurring operational cost, the staff are being paid, and yet no science is being done.

As a result, it can mimic the heroic nature of getting a game released, where the deadline is all-important, and shortcuts are taken without thought for the future. As the project entered operations, however, the pace of change that had previously been tolerable in commissioning now became burdensome, since software changes had large downstream impacts (e.g., on data quality assessment).

Despite the challenges, this billion-dollar telescope is now conducting groundbreaking science, backed by powerful software.

Note

1. This extends to other scientific applications as well; the need to process large volumes of data for scientists is not unique to astronomy.

Further Reading

The Atacama Large Millimeter Array website is at: https://www.almaobservatory.org. The software packages can be found at CASA: https://casa.nrao.edu. Management and requirements for next versions of ALMA can be found at: https://www.almaobservatory.org/en/about-alma-at-first-glance/the-people/the-alma-board/the-alma-board-meetings-summaries/.

The papers this case study is based on are:

A. M. Chavan, B. E. Glendenning, J. Ibsen, J. Kern, G. Kosugi, G. Raffi, E. Schmid, J. Schwarz, "The Last Mile of the ALMA Software Development: Lessons Learned," Proc. SPIE 8451, Software and Cyberinfrastructure for Astronomy II, 84510Q (September 24, 2012); doi: 10.1117/12.925961.

Ralph Marson, Rafael Hiriart, "The Transition from Construction to Operations on the ALMA Control Software," Proc. SPIE 9913, Software and Cyberinfrastructure for Astronomy IV, 991304 (July 26, 2016); doi: 10.1117/12.2233584.

B. E. Glendenning, G. Raffi, "The ALMA Computing Project: Initial Commissioning," Proc. SPIE 7019, Advanced Software and Control for Astronomy II, 701902 (14 July 2008); doi: 10.1117/12.787569.

SKA's Tango choices are documented in L. Pivetta et al., "The Ska Telescope Control System Guidelines and Architecture," 16th Int. Conf. on Accelerator and Large Experimental Control Systems, Barcelona, Spain, 2017, https://accelconf.web.cern.ch/icalepcs2017/papers/mobpl03.pdf.

9 Technical Debt in Machine Learning Systems

—with Humberto Cervantes

> Essentially, all models are wrong, but some are useful.
>
> —George E. P. Box

While technical debt is a concept most often applied to the vast amount of traditional software programs (such as those written in Java, C++, COBOL, and FORTRAN), today's software systems increasingly include probabilistic data analysis components that rely on patterns and inference to make decisions. These are *machine learning systems,* and the techniques and skills for building and maintaining them are different. It is likely that you have used such a system or perhaps maintained one: online ad networks have machine learning components, as do text autocorrection, facial recognition in image libraries, and banking mortgage applications.

Not surprisingly, technical debt is an issue in these systems as well, and there are some forms of debt that are specific to machine learning systems. In this chapter, we first discuss the differences between traditional and machine learning systems and then describe ways in which technical debt in machine learning systems can be identified, managed, and ultimately, avoided.

Machine learning systems have the same technical debt risks as traditional software—machine learning components are just software, after all. They have external dependencies on libraries, data processing code, computational algebra, documentation, and so on. Chapters 3–8 apply to these systems, too. They also introduce new types of technical debt centered around design choices, integration, explainability, configuration, and testing. We focus on these new types of debt in this chapter.

9.1 Background

The traditional approach to solving problems has been to write algorithms and data processing code to capture problem reasoning and states. We call these *software-intensive systems*. For instance, let's say our system is a robot, and we want to program the robot to navigate a hallway. We might create a planner and feed the robot data, plans, and goals, and let it process data about the state of the world. The robot's software would then use the planner in a loop to make decisions: if it sees some trigger, then it goes left, else right. This has been called a symbolic approach to computing: symbols represent the system's knowledge of the world, such as the robot's position, whether a door is open or closed, and the robot's eventual goals (get to end of hallway).

Increasingly, though, we are creating and operating software to work in a new, probabilistic world. We don't need to program *all* behaviors into a system. There are some classes of behaviors that the system can simply learn (or relearn as the data and environment change) using powerful learning algorithms and sufficient amounts of suitable training data. These components of the system are black boxes that we *train* more than we program. We call these "black boxes" because the reason why a given output was produced is often impossible to trace and reason about. To train them we just need to identify the data sources, create a rough skeleton of the learner, and then allow the power of numerical optimization to find the best set of features and weights to solve the problem. Now a robot can be told merely to get to some destination and to learn as it goes. These are sometimes called *data-intensive systems*. In this chapter we refer to them as *machine learning systems*.

The advantage is that programming a robot with the data-intensive, probabilistic approach opens up new possibilities that are not amenable to traditional rule-oriented programming techniques. We can typically leverage preexisting libraries and architectures (such as types of learning approaches, like neural networks, decision trees, etc.). The engineering effort focuses on preparing and labeling data, selecting important features, and evaluating algorithm performance, rather than writing, testing, and debugging code.

Moreover, the black box of the learning component continues to improve as more data becomes available—for example, as our robot traverses new, unknown terrain. This requires no human "in the loop" of the control structure. As Andrei Karpathy wrote, "a large portion of real-world problems have the property that it is significantly easier to collect the data (or more generally, identify a desirable behavior) than to explicitly write the program."

Clearly the new machine learning-intensive components of software systems do not replace traditional software systems. Instead, they coexist. Think about a payroll system. Inside this system there is, and always will be, a large amount of traditional software, doing database queries, updating records, calculating payments, issuing checks, and interacting with third-party libraries and external systems. But now, there will also be machine learning, probabilistic black box components such as the fraud identification component, the benefits optimizer, or the hiring optimizer. These machine learning black boxes integrate into the rest of the software: they receive data inputs (like current deductions from payroll) and produce insights in the form of actionable metrics or recommendations. For example, in a payroll system the machine learning fraud detection component might highlight employees or specific benefit claims that are deemed likely to be fraudulent.

One of the difficulties with machine learning systems is that, as opposed to traditional systems that have failures and stop functioning, such as null pointer exceptions, machine learning systems will usually continue to function, but do so inadequately. For example, if the characteristics of the data that is used for prediction begin to differ from the ones that were used for training, overall accuracy will start to degrade. The system has not failed in the sense that the execution has stopped, but the results it is providing are nonetheless incorrect. For example, consumers may get less useful recommendations for what to buy, reducing sales. This reduced accuracy is potentially harder to detect than a traditional system that is malfunctioning.

In this chapter we elaborate more on the risks that each of these pose to systems that are increasingly machine learning hybrids. Keep in mind that our advice in the other chapters still applies here. For example, social debt (chapter 10) is a potentially greater risk, since two new teams—with very different mindsets and cultures—are now involved with the system: the data engineers and the machine learning engineers. This is also an emerging area, since machine learning hybrids are relatively new to production software environments. As a result, we do not claim our list is exhaustive.

9.2 Identifying Machine Learning Debt

There are four main types of machine learning debt to be aware of. These stem from design choices revolving around integration, explainability, configuration, and testing.

9.2.1 Design Choices

Developing a machine learning system involves a significant effort in the choice and development of the model and associated algorithms that are used to make predictions. There are two important activities that are performed to achieve this goal: one is related to data, and it involves the selection and transformation of input data into *features*. The second involves the selection and training of an algorithm that uses these features to make predictions. When performing these activities, many design choices are made. These choices introduce many opportunities for taking shortcuts or making incorrect design decisions that translate into debt. We will now introduce two cases where poor design decisions introduced substantial (and avoidable) technical debt.

Case 1: In this project a talented but inexperienced data scientist spent six months in the development of a learning model which, once in production, did not produce high quality results. A group of consultants analyzed the situation and concluded that the data scientist had not spent enough time performing data analysis to understand the correlations between features to identify their importance. As a result, they made an incorrect design choice—the selection of the paradigm, in this case reinforcement learning—and spent (wasted) a considerable amount of time fine-tuning this approach. After doing a more thorough analysis of the data, the consultants concluded that a different learning paradigm would have been more appropriate and that the work that had been done previously could not be leveraged and had to be restarted from scratch. Not dedicating sufficient time to data analysis and quickly jumping to the selection of an algorithm and its fine tuning is a common, and costly, design mistake. Instead, it is necessary to dedicate sufficient time to analyze the data and understand aspects such as correlations to choose and tune an appropriate model.

Case 2: Other costly design choices may occur after the model has been developed and once it is put into production. For instance, not having carefully considered where the model executes in the production environment may require a full rewrite of the model. An example of this was observed in a project where deep learning models were to be used on edge devices in an industrial production facility for processing images. Initially the model development team simply assumed that the Python algorithms they had developed could be run in production by using GPUs (powerful graphical processing units, specialized for numerical computing) as edge devices. While this allowed the model to be put into production quickly, it had major disadvantages: the GPUs have a short lifespan in practice (around two years) after which they need to be replaced, and this involves stopping the production line. A proposed solution was to use FPGAs (field programmable gate arrays) instead of GPUs on the edge, the benefit being that FPGAs have a lifespan of about ten years. However, the use of FPGAs involves a complete rethinking of the architecture of the neural networks, and that is essentially a different model. The original pipeline, which employed the neural networks using Python, was simply not directly portable to FPGAs, and so the pipeline also needed to be reworked. These costly design decisions could have been avoided if the decision to use FPGAs had been considered from the beginning.

Some useful metrics for assessing the overall state of technical debt in your machine learning system include evaluating the differences in performance and accuracy between the model being used and state-of-the-art models for which benchmarks are available. This metric, if it had been employed in case 1, could have identified the poor model choice far earlier in the project's development. Other metrics can be helpful to evaluate the technical debt in the infrastructure established to support the development of the system. For instance, it can be useful to measure how frequently updates to the model can be performed and what is the level of automation that has been established to support retraining and updating of a model that is in production. Low levels of automation and long or costly update processes that require human intervention or changes in physical infrastructure may reveal potential debt areas. This metric, if it had been employed in case 2, could have identified the dependency on the GPUs as a debt red flag.

9.2.2 Integration

Integration debt is the kind of technical debt that comes from poorly thought out integration dependencies with other, nonmachine learning components, such as data processing components. Since machine learning components are not well understood by other, nonmachine learning teams, they must be carefully integrated with the rest of the system. Now, one might argue that since a machine learning system is a black box, it must be highly modular, and therefore it should integrate easily. In this sense, the machine learning component is no different than any other component: we should document the expected inputs and outputs. In another view, however, the machine learning component is in fact introducing an entirely separate ecosystem of libraries and dependencies, which other teams may not know or have expertise with. For example, figure 9.1 shows an example, inspired by a paper from D. Sculley, of the ecosystem that a machine learning component—a model—relies on.

In addition to the mechanics of introducing the component itself and wiring it together (e.g., via a data flow structure), there are two less obvious sources of possible debt. First, data dependencies, such as entanglement

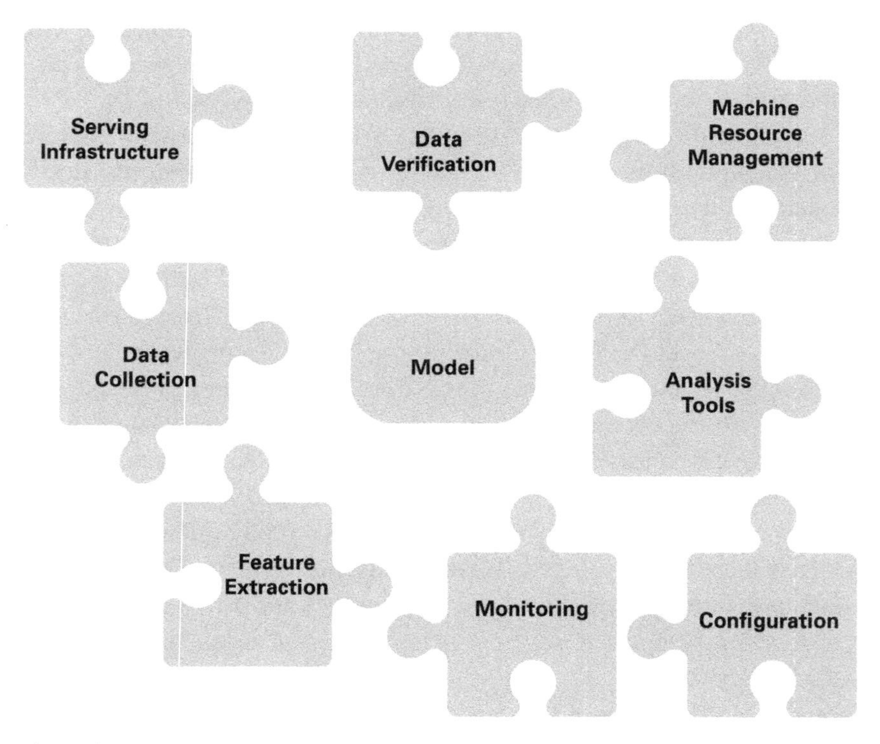

Figure 9.1
The machine learning ecosystem. Note the small size of the machine learning code—the model—relative to the entire system.

and hidden feedback loops, mean the machine learning system is likely to be tightly coupled to the sources of the data.

Second, social debt is likely to increase, since the teams that maintain the different parts of the new system are likely different (the machine learning team, the data team, the UI team, the workflow team). Using a cross-functional team structure can help, but in the near future upskilling existing developers in the world of machine learning software is more important. This will provide the necessary knowledge to at least understand where they should seek expert help.

Finally, integration is likely to cause tooling and delivery problems, since tools for creating, editing, and debugging machine learning components are still emerging and changing rapidly. The equivalent to the IDE for machine learning systems is yet to be determined. Some candidates are computational notebooks like Jupyter, but it is possible existing IDEs will simply upgrade to support these new paradigms. For example, debugging a machine learning system often involves extensive data analysis and model validation rather than breakpoints and memory inspection.

To identify integration debt, we recommend measuring the number of new libraries and components introduced over time on a per-release basis or at some other important milestone. A high-level, cross-project review board is essential to identify existing version and dependency incompatibilities. Specific labels in the issue tracker can also be used to highlight data exchange issues.

9.2.3 Explainability

Depending on their complexity, some machine learning components can be almost impossible for a human to interpret and fully understand and are thus seen as black boxes. Think back to the robot at the beginning of the chapter. The rationale for the route taken in the traditional software system is obvious, and we can debug it by asking the robot to trace the software choices made (e.g., each branch through the code). But in machine learning-intensive software systems, understanding these choices is much more difficult. Each choice the robot makes is the result of nothing more than a set of weights on a network architecture. While tools exist to give some probabilistic path through the network, this may not be sufficient in high criticality systems. For safety-critical code, for instance, it is extremely important to know why a given action happened.

Explainability is a quality attribute for machine learning systems that refers to a clear and obvious connection between a machine learning label and the internal mathematics that produced that label. For example, if we have a system that labels job applicants as "hire/no-hire," we may have to explain (to ourselves, or to regulators) how those labels were derived. With complex models, this may be very difficult or impossible. One aspect of this explainability is the design decision that led to a particular model being used.

Technical debt, in the form of documentation debt (see chapter 8), can be introduced if a particular machine learning model is selected and trained, but no clear rationale for this choice is documented. Documentation debt can also be introduced if the model selected does not support explainability, which may later be needed (e.g., a court case requires exposing how a decision was made). The team would have to rework the approach to add back the missing ability to clearly explain decisions, for example, by using a different algorithm. A potential form of technical debt in machine learning systems is to optimize purely for performance of the model, without considering explainability. Like any technical debt, this shortcut may lead to longer term problems and higher costs.

Many outages in traditional software systems are traceable to strange edge cases. In probabilistic systems, something equivalent occurs when the model in production is confronted with data which was not available at the time of training. Consider, for example, a system that does product quality control using computer vision methods and machine learning for image preprocessing. A new product that uses different materials from the ones used to train the model will probably result in the model not being capable of working effectively. A good way to identify these risks is to look for the presence of model documentation in the existing system. Can the maintainers of these models point to a document that justifies the current production machine learning choices? This is also a useful exercise to conduct with product/project managers or other customer-facing staff. Given a decision suggested by the software, can the rationale for that decision be traced back to its origin? Keep in mind that requirements traceability is often hard to do even with nonmachine learning approaches (as we discussed in chapter 3).

9.2.4 Configuration

Configuration of the machine learning system refers to the various parameter choices that come with machine learning approaches. Technical debt is introduced when we are not clear why configuration choices of the model were made. For example, gradient descent algorithms typically have a learning rate parameter. Too fast, and the true optimum might be missed; too slow, and the model takes too long to converge. There are many ways to choose this parameter. Technical debt can occur if the choice is not made rationally (e.g., by following current best practices for tuning), and more importantly, if the choice is not captured somewhere—ideally, as we discussed in chapter 8 (documentation debt), close to the production code. Jupyter IDEs (computational notebooks) can interweave code and text nicely, but the risk is that porting the code to production systems might lose the associated text-based documentation.

Similar configuration risks exist in many other facets of the machine learning system. For example, we will need to capture the choices for the underlying neural network architecture or machine learning algorithm (e.g., support vector machines, four-layer recurrent neural network, etc.) We will also need to document how we did (or did not) do hyperparameter tuning and model validation steps. In general, employing poor data science practices (e.g., failing to split datasets into test/train/validate subsets) has the potential to introduce technical debt. To identify this type of technical debt, not only ensure that documentation exists but also survey staff to understand current practices and training levels. In many systems, configuration code is just as complex, if not more complex, than the actual machine learning components. Bad practices in managing these options can also cause configuration technical debt.

9.2.5 Testing

Testing probabilistic systems is different than testing more traditional software systems. For example, consider a web app that asks for age on sign-up. The web app stores this value in a data store, and in a traditional codebase we would likely test to ensure that, for example, age was not negative, age was less than 130, and other business-specific concerns (perhaps cases where age >18). If the values for that field were generated by a machine learning approach, for example, based on friend networks or shopping habits, the field that was an integer now comes with a confidence measure. We can no longer assume that if age > 18 the user is an adult: perhaps someone else was using the account, or they share a credit card with a parent. Our tests now must take this probability into account as well.

There is other technical debt possibly related to testing, and the paper by Breck and colleagues (see the Further Reading section) gives an expansive list of tests that can be applied to production machine learning systems based on work at Google. They focus on reliability and suggest tests to determine how informative features are (to avoid wasting computational resources on uninformative features) and how to assure that higher level requirements exist and are met for machine learning systems (suggesting that some degree of traceability remains important in these systems). Note that this testing effort is above and beyond any testing of the traditional software in the system (as discussed in chapter 6). A simple measure for this risk is the degree of test code that exists for each version of a machine-learning model—for instance, the test coverage of the model. This indicator should assess the tests of the model itself (i.e., accuracy, false-positive rate, or problem-specific indicators) as well as tests of the machine learning infrastructure.

9.3 Managing Machine Learning Debt

There are a number of remediations we suggest to manage the debt in machine learning systems. We point out that there is a fundamental tension between *model development* and *model deployment and serving*. That is, when developing models, we do scientific experimentation to find the right combination of features, architectures, and parameters. This experimentation is inherently about discovery and stands in opposition to engineering, which is what we do more often in software-intensive systems. In deployment and serving approaches, we are more concerned with stability and maintainability. Failures are bad and have serious consequences. "Failure" here has a more subtle meaning than in most traditional software: a failure can be experienced when a machine learning system is not producing the correct results, even though it is still executing. Failures in experimental settings, however, are encouraged and help us learn. Properly managing technical debt in machine learning systems means understanding what context a particular system is operating in. A lack of reliability tests might be less of a concern for experimental projects, whereas they would be unacceptable in production software.

We point out some possible remediations to technical debt in machine learning systems: use simpler algorithms, version scientific notebooks, and create cross-functional teams.

9.3.1 Simpler Approaches

While there is a temptation to work with the most complicated model, drawn from the cutting edge of research, such complexity of course brings with it a greater potential for technical debt. The problem is that more complex models and cutting-edge approaches, such as the latest word embeddings, appeal to engineers and data scientists, but the business value of these approaches is not always obvious. For example, new approaches may significantly increase training time but have little impact on business value. Good management approaches to technical debt will always compare the more complex approach with a more stable, simpler baseline. For instance, logistic regression or support vector machines might perform nearly as well as a complex deep neural network approach and are much easier to explain and maintain. So you should always start with the simplest approach unless there is a solid reason, backed up by experience and ideally empirical data, to adopt a more complex one. Movement to a more complex approach should only be catalyzed by inadequate performance of the simpler approach.

9.3.2 Versioning in Exploration

While above we said that some software practices might be less important in an exploration context, some degree of versioning is important everywhere. For one thing, comparing multiple approaches to the problem is only possible if you have a record of what approaches were used and where they live. Version control tools that work well with software code, however, may not be the best approach for experimentation. There are custom machine learning workflow tools such as DVC (Data Version Control[1]) or Netflix's Metaflow that are able to store input data, configuration, and the code so that all of the information needed for reproducibility is stored in one place, for easy access and comparison.

A lot of machine learning technical debt comes from lack of reproducibility. As figure 9.1 shows, configuring machine learning systems is complex and nontrivial. Since the field is moving very quickly, one way of managing your technical debt is by insisting that best practices in documentation are followed. It is common for model parameters to be scattered across model code. For example, when creating a machine learning model you need to choose the learning rate (for a neural network) or the value of k (for k nearest neighbors). These are nothing more than the magic numbers code smell that we have known about for decades. Move these magic numbers into a configuration file, along with the rationale for the choices that you have made, so that you can offload specific configurations as separate experiments. This will allow you to keep track of what was run and when. It has the nice side effect of keeping the code cleaner too, as all the actual machine learning code needs to do to run is load a config file that contains layer sizes, learning rate, and other important parameters.

9.3.3 Cross-Functional Teams

Our final management tip relates to chapter 10 and the importance of teams and the risk of social debt. Since machine learning is really a new set of specialized skills and knowledge, this increases the risk of communication challenges and other community smells. Machine learning activities typically involve specialists in data engineering (to manage data pipelines and storage), data scientists (to run experiments), and machine learning theorists (to work with the models themselves). Cross-functional teams are one way to reduce the risk that silos or other social smells emerge between these groups.

A cross-functional team is one that is organized around a business goal or feature rather than a technical capability. This is also known as vertical slicing, as the team's mandate is to improve customer retention, or manage user authentication, rather than simply having the database team and the UI/UX team. The risk of this organizational style is that each role within the cross-functional team might feel disconnected from their peers, and knowledge transfer might be reduced. But these risks can be addressed with some version of a matrix organization or simply arranging regular opportunities for cross-team communication and knowledge sharing. One strategy for achieving this is to ensure that some employees are sufficiently cross-trained so that they can take on the role of boundary spanner (as discussed in chapter 10). For example, it is recommended that a machine learning engineer already be a software engineer, who has been trained in machine learning tools and techniques. This machine learning engineer is ideally poised to be a boundary spanner.

9.4 Avoiding Machine Learning Debt

The best way to avoid technical debt in machine learning systems is to focus on how to properly design machine learning systems. There are two major sets of design decisions related to creating and fielding these relatively new pieces of software. One is the architectural considerations involved in creating the initial machine learning model, and the second is the set of considerations involved in putting that model into production. We now deal briefly with each of these concerns.

The architectural concerns to support the model-development and refinement part of the machine learning system lifecycle revolve around:

1. The ingestion, cleansing, and transformation of training data, including labeling the data to support supervised learning
2. Feature engineering—the selection of specific properties of the data that are of interest for learning
3. Model selection, where a machine learning algorithm is chosen, parameterized, trained, and tuned
4. Model persistence, where the model is prepared to be transformed to a production environment

The architectural concerns to support model deployment and serving revolve around:

1. Deploying the model to a serving environment that may have different characteristics than the training environment
2. Ingestion of new (production) data
3. Data transformation, cleansing, and validation
4. Prediction (potentially including retraining), which is the reason we built the machine learning system in the first place
5. Serving of results to other systems or to end-users

To avoid introducing unwanted technical debt in each of these processing phases we must be conscious of the concerns of each phase. We will first consider the concerns addressed when developing and refining a model. First, you must consider the tools you will use to ingest, store, and transform the data—perhaps employing a data broker and a data transformation pipeline. For feature engineering you need to decide how to split data into subsets for training, testing, and validation.

These datasets are potentially large, and they need to be logged and versioned for testing and ensuring reproducibility. In certain domains, the size of data sets can be considerable. For instance, genomic data sets may be so large that loading the complete dataset could take weeks. Attention needs to be put in the infrastructure to avoid loading the entire file, for example by using specialized batch loading techniques. As some training data may be sensitive in nature, you need to think about how to deal with security and privacy concerns. When selecting a model you will need to think not only about its accuracy but also its scalability characteristics and its availability on your production platform. Finally, these models can be quite large, and so there may be critical design decisions that you need to make around model persistence, serialization, and optimization.

Now we move to the concerns revolving around moving your machine learning system into a production environment. The first thing to do is to make the trained model available in a production environment, and so this immediately raises concerns about the transfer mechanism, about versioning, update, and rollback, and about how the model will be deployed (e.g., blue-green deployment, canary deployment, etc. See chapter 7 on deployment debt for more details).

Once this new model is in place in production it must be fed data. You need to decide how this data will be collected, ingested, and transformed. At this point you are able to extract features from the data, ideally following the process established in the model-development phase. But what if the characteristics or distribution of production data or features differs from that seen in model-development? In that case model retraining might be necessary. To assess this, you will need to establish, collect, store, and analyze a set of validation outputs.

Finally, predictions are made by the system in production. These predictions should be monitored, and if these predictions do not meet your quality criteria, retraining might again be warranted. In this case you need to decide if you want local or remote retraining—or perhaps a hybrid model. Local retraining has the advantage that it does not require additional network bandwidth. It is limited to minor changes in data, however, and client devices typically have less processing power than servers. This might limit the kinds of processing and retraining that can be accomplished. At the extreme, as described in case 1, the entire model may need to be replaced. As mentioned above, this kind of debt can be avoided by evaluating the differences in performance and accuracy between the model being used and state-of-the-art models for which benchmarks are available.

As you can see, there are many decisions that need to be made at each stage of the training and serving pipelines. However, a robust consideration of these issues will help you to avoid unsustainable levels of technical debt in your machine learning systems.

9.5 Summary

Martin Zinkevich wrote the following: "Even with all the resources of a great machine learning expert, most of the gains come from great features, not great machine learning algorithms." A lot of the technical debt thus comes in the form of forgetting that machine learning systems are no different than other software approaches: they are there to solve problems the organization has. Many of the chapters in this book are therefore equally applicable to data-intensive machine learning systems as they are to traditional software-intensive systems:

1. Like software-intensive systems, data-intensive systems are still deployed with business or organizational objectives to satisfy. Therefore, tracing from requirements to deployed models is very important.
2. Design and architecture must take into account the special nature of how machine learning systems work and the new dependencies that therefore arise.
3. It is still possible to take shortcuts, but the shortcuts mostly come from doing poor data science. Shortcuts in data science practices might involve, as an example, failing to validate that the production data is statistically similar to the training data.
4. Testing is, if anything, more important, as testing machine learning systems has a much shorter history and so our tools and techniques for testing are less mature.
5. Deployment debt can occur if operations teams are not clear on the new demands that deploying machine learning systems involves, such as the nature of the model being transferred, the mapping of that model onto the production hardware, and the differences between training data and production data.
6. Like all software, documentation that properly captures why and how decisions were made is vital.

One thing we have noticed is that many times technical debt is introduced unknowingly, often because the developers had no idea that their code would become production-critical. We suspect the same is true for machine learning models: a model will sometimes be put into production because it worked, and twenty years later, it will be a critical part of the organization's IT systems. Take care now to ensure that twenty years from now the machine learning code will be maintainable and clear.

Note

1. https://dvc.org/.

Further Reading

The first reference to technical debt in machine learning systems is from D. Sculley, Gary Holt, Daniel Golovin, Eugene Davydov, Todd Phillips, Dietmar Ebner, Vinay Chaudhary, Michael Young, "Machine Learning: The High-Interest Credit Card of Technical Debt," workshop at NeurIPS, 2015.

Eric Breck, Shanging Cai, Eric Nielsen, Michael Salib, and D. Sculley wrote about testing production machine learning systems in "The ML Test Score: A Rubric for ML Production Readiness and Technical Debt Reduction," *Proceedings of IEEE Big Data*, 2017, https://ai.google/research/pubs/pub46555; and there are numerous blog posts and online courses in creating machine learning models. https://www.fast.ai is one specifically designed for software engineers.

Martin Zinkevich authored Google's rules for machine learning (from which the quote is taken). Last updated in 2019, they can be found here: https://developers.google.com/machine-learning/guides/rules-of-ml/. They are forty-three useful rules for building production-level machine learning systems, based on Google's experiences with ads and social networks.

Andrej Karpathy first identified challenges with "Software 2.0," as he terms it, in his 2017 blog post of the same name: https://medium.com/@karpathy/software-2-0-a64152b37c35.

The AllenAI institute and its researcher/developers have been at the forefront of well-engineering data science workflows. They describe some of this in their tutorial, available at: https://github.com/allenai/writing-code-for-nlp-research-emnlp2018/blob/master/writing_code_for_nlp_research.pdf.

10 Team Management and Social Debt

—with Damian A. Tamburri

> Debt is a social and ideological construct, not a simple economic fact.
> —Noam Chomsky

It is very rare that projects fail because of purely technical reasons. The vast majority of failures are, at least in part, due to communication or management issues. The structure of an organization has a direct effect on the organization of a project, and therefore on its source code hierarchy or even testing process. Communication, management, or team organizations can have drastic effects on technical debt in software artifacts. This chapter explores the relation between organizational context and technical debt. Social debt exists when organizational problems burden software development. Social debt exists when the organizational structure is incongruent with the structure of system artifacts. This incongruence invariably leads to the same kinds of problems as technical debt in other artifacts, including lower code quality, loss of agility, and high downstream costs. In this chapter, we first introduce the concept of social debt. Then our chapter follows our conventional flow, explaining how to identify it, how to manage it, and how to avoid it.

10.1 Defining Social Debt

While software engineering has traditionally focused on artifacts—code, processes, tools, and so forth—just as important are the people and their teams, their structure and characteristics, and all that is behind the humans that actually build those artifacts. The impetus for focusing on people can be traced back to the origins of software engineering as a discipline. Melvin Conway described in 1968 what has now become known as Conway's law, stating that the structure of a software system reflects the social structure of the organization that produced it. Conway cited several pieces of anecdotal evidence to support his claim; for example, a COBOL compiler that was developed by five people and ran in five phases, and an ALGOL compiler that was developed by three people and ran in three phases. This notion—that the social structure of an organization matters and has profound implications in practice—has been shown over and over in the more than fifty years since Conway first wrote this article.

Since there are countless ways to structure organizations and countless ways to structure systems, it stands to reason that some of these structures are suboptimal. We define social debt as the mismatch between system structure and organizational structure. Furthermore, it stands to reason that some mappings of organizational designs onto system designs will also be suboptimal.

To give a brief, anecdotal example, consider that you have three teams that, collectively, are working on a software project, perhaps an e-commerce system. Each team is in charge of a major subsystem. If this e-commerce system is broken up into the typical three-tier architecture—back-end tier, business-logic tier, and front-end tier—then one would expect, for example, that the database team would need to have greater interaction with the business-logic team than with the user-interaction team. Furthermore, one might expect substantial interaction among the teams in adjacent tiers; for example, as the business logic changed this would likely need to be reflected in user-interface changes. If the teams did not communicate in this way—perhaps the user-interface team spoke seldom to any of the other teams and only responded slowly and reluctantly to requests for changes or for information—then this would be a form of *social* debt. This kind of miscommunication is not uncommon, particularly in distributed projects and in projects where different components are developed by different teams.

Social debt is the additional project cost connected to all kinds of socio-technical incongruences. For example, think of silo effects, where a project is grouped into organizational silos and these silos have little interaction with each other. If those organizational silos create artifacts that depend on each other, then this lack of personal interaction will likely lead to problems. For example, one team might not document everything needed by the other team, assuming that such documentation is not needed. Or if one team is not aware of a problem another team is facing (e.g., some performance issue), they might develop their own optimization (such as multithreading), which might create a new problem (e.g., data consistency if the initial code is not thread-safe).

Ideally you want sociotechnical *congruence* in your project: components that are tightly connected should be developed by teams that are in constant communication with each other, ideally co-located. Components that have little to no interaction can be developed by teams that are similarly isolated from each other. In both of these cases we have congruence between the social structures and the technical structures.

So how do we identify social debt (and its impact on technical debt), how do we manage it once we know it is there, and how do we attempt to avoid it in the first place? These are the topics of the next three sections.

10.2 Identifying Social Debt

Similar to the use of code smells to identify coding issues, we use the concept of community smells to detect and mitigate suboptimal social structures and behaviors. An organizational structure is a network of people (and artifacts, such as code, tests, issues) that is focused on achieving some collective goal (such as delivering a piece of working code). We can model these organizational structures as networks. Over the past decade there has been increasing interest in social networks—networks of nodes (typically some kind of actors, people, or even organizations) and links between the nodes (relationships)—to model organizational structures. The field of social network analysis has existed for well over a century, but it has grown tremendously in importance with the ubiquity of computer-mediated social networks.

Social network analysis allows us to gain insight into the connections within the network, the distributions of nodes and edges across the network, and how the network is segmented into subgroups such as cliques. Social network analysis can determine, for example, when networks, or portions of them, are balanced or unbalanced (e.g., if one group is primarily sending messages and another receiving them), whether they are centralized or distributed, whether they have a high or sparse density of interconnection, and so forth.

We are all part of networks: at home, in our social lives, and in our business lives. If you think about how social networks are structured, and if you think about this in the context of a business, some forms of network organization are going to be suboptimal.

This is not a problem for small teams, of three or five people, as depicted in figure 10.1. But if you are part of a team of just seven people and everyone

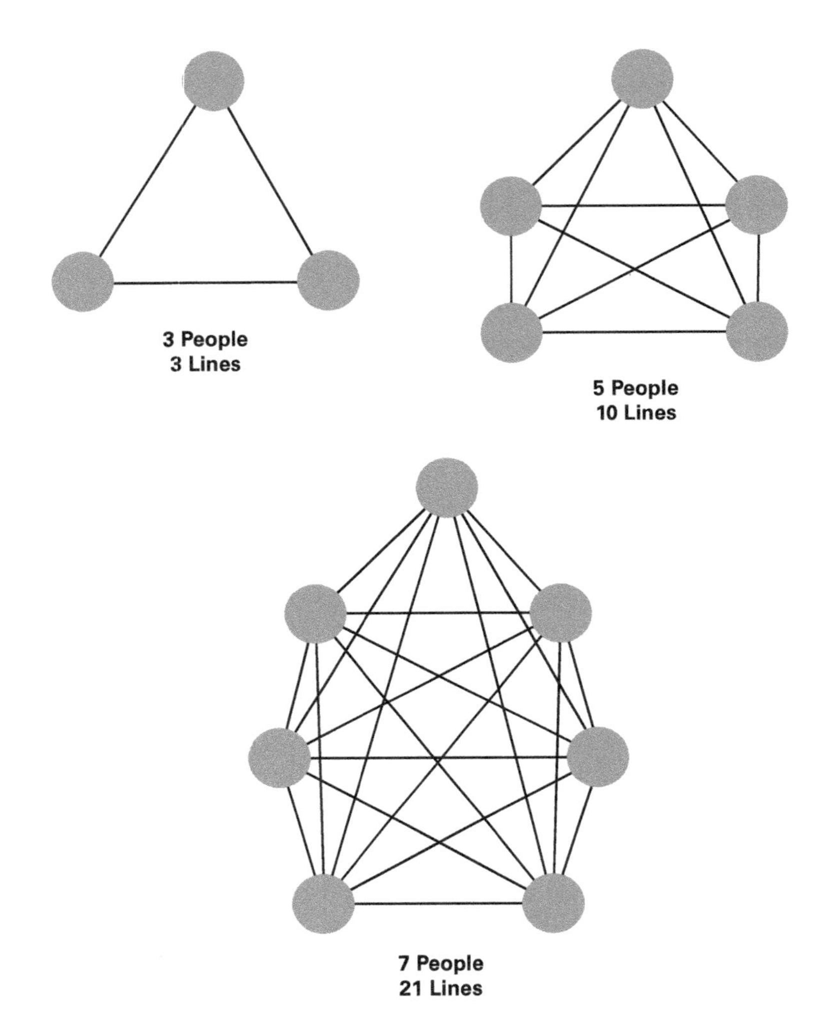

Figure 10.1
Exponential growth in communication networks.

is expected to be informed of the activities and decisions of everyone else, then communication is going to be rather complex and time-consuming. What if your team has twelve people? Again, if this team needs to be strongly interconnected, then clearly no actual work will get done because working time will be preempted by communication. This team of twelve has the potential to generate sixty-six different lines of communication. Well, what if we divide those same twelve people into three teams of four people each? If these subteams are strongly interconnected, then nothing is gained; but if the teams have very little communication with each other, then there is a high likelihood that some of the decisions made in one subteam will have an impact on one or more of the other subteams and that those decisions will be inadequately communicated.

Most organizations can and will obviously do better than this naive model. The point of this simple example is to get across the idea that social networks matter and that not all network structures are going to work efficiently. To identify social debt, therefore, we need to think about *which* patterns of social interaction are harmful.

We cannot think about the social interaction in isolation. Social interaction needs to be considered in a context; that is, in light of the mapping of the social network onto the technical network—for example, onto the artifacts like source code that the teams work with.

Let's consider a simple example. Team 1 and team 2 are working on software components A and B respectively, and components A and B have high coupling—creating a technical network. Perhaps they are integrated together, interoperate with each other, share common resources, or there is a timing dependency between them. It would be natural for team 1 and team 2 to have high-bandwidth communications with shared meetings, chat groups with both teams, and/or regular meetings between tech leads of both teams. In other words, the teams should respect Conway's law, and the high coupling of their respective components should be mirrored by a high level of communication, interaction, negotiation, and so forth. So we need to be concerned with the mapping of any social network onto a network of artifacts, which we call a *technical* network. The interesting point to consider here is that the structure of the social network can influence the structure of the technical network. How our social network is managed and operates is naturally an important consideration.

There are a set of common community smells that are found repeatedly on software projects. These smells are listed in table 10.1. They lead to undesirable consequences such as low software quality. For example, we see the phenomenon of free-riding occur often in smelly communities. Free-riders are those who benefit from some common good without contributing to it. Clearly you would want to detect and ideally eliminate such smells before they cause real trouble.

These community smells, along with their contexts (i.e., which community types are likely to develop which conditions), causes, and consequences that may incur technical debt are listed in table 10.1.

Many large projects are subject to a wide diversity of experience. That is, they have developers with greatly varying levels of competence and experience. This can lead to the cognitive distance, time warp, and unlearning smells. Let us consider cognitive distance—the perceived distance between developers and their peers with considerable differences in their technical, educational, and even social backgrounds. If left unaddressed, this smell can lead to wasted time and resources, and even the division of a project into clans that are mutually distrustful of each other. Having a team consisting only of people with identical backgrounds is not desirable either, as many studies show diversity in background is valuable for creativity, among other benefits. In such a case, you might end up with people lacking some skills and making mistakes. For example, if you have an isolated team with no frontend experience, the first user interface they will produce will be suboptimal. Having more diverse skills (or just having communication with other teams) would help this team to find relevant resources to deliver a better interface faster. Building team cultures with common standards and practices, and on-the-job training, are some ways to mitigate this smell.

To take another example, excessively high *power distance*—the distance that less powerful members of a community perceive between themselves and with power-holders (such as architects)—can negatively affect trust and productivity in teams. For example, some very experienced developers will not trust more junior developers, who will then be afraid to contribute or share their ideas. Another case is when developers develop imposter syndrome and avoid contributing, fearing their opinion is not good enough for the team to be shared.

These smells are quite common. In one study of ninety practitioners from nine separate organizations (covering the domains of aerospace,

Table 10.1

Community smells (adapted from [Tamburri 2016] and [Tamburri 2019])

Smell	Description	Cause	Consequences
Cookbook Development	Developers who are stuck in their ways and refuse innovative ideas or ways of working (e.g., agile methods or DevOps)	Thinking in an old framework—for example, the waterfall model	• Mismatched expectations between the customer and the rest of the community
Time Warp	A change in organizational structure and process that leads people to wrongly assume that communication will take less time and that explicit additional coordination isn't needed	Experience diversity	• Low software architecture quality • Malfunctioning software or code smells • Losing face in the community • Unsolved operations issues • Dissatisfied customers
Cognitive Distance	The distance developers perceive on the physical, technical, social, and cultural levels regarding peers with considerable background differences	Experience diversity	• Wasted time • Pitting newbies versus experts • Faulty or smelly code • Additional development costs • Wasted operational resources • Lack of optimal understanding across different operational areas • Mistrust across the development network • Misinterpretation of expectations
Newbie Free-Riding	Newcomers being left to themselves regarding understanding what to do and for whom, with the consequent free-riding of older employees (that is, the economic free-rider problem applied to software engineering)	Unknown	• High work pressure • Irritation • Demotivation of less senior members

(continued)

Table 10.1 (continued)

Community smells (adapted from [Tamburri 2016] and [Tamburri 2019])

Smell	Description	Cause	Consequences
Power Distance	The extent to which less powerful members of a software development community perceive or accept that others have more power	Lack of architecture knowledge sharing	• Additional project costs • Financial loss • Lost bids
Disengagement	Thinking the product is mature enough and sending it to operations even though it might not be ready	Lack of engagement in development Lack of curiosity	• Missing software development contextual information • Wild assumptions • Increased QA costs • Loss of customer trust
Priggish Members	Demanding of others (e.g., operations) pointlessly precise conformity or exaggerated propriety, often in a self-righteous or irritating manner	Unknown	• Additional project costs • Frustrated team members • Negative impact on team spirit
Institutional Isomorphism	The similarity of the processes or structure of one subcommunity to those of another, whether the result of imitation or independent development under similar constraints	Excessive conformity to standards Lack of innovation Using formal structures to achieve community goals Rigid thinking	• A negative impact on team spirit • A less flexible or static product • Lack of innovation • Stagnation • Lack of communication or collaboration
Hypercommunity	A hyperconnected community that's susceptible to group-think and influences its subcommunities	Unknown	• Increased turbulence • Buggy software • Missing opportunities from less influential members

DevOps Clash	Clashes in the mix between development and operations from multiple geographical locations, with contractual obligations to either development or operations	Geographic dispersion	• Increased project costs • Lack of trust-building • The inability to bridge between different thought worlds across development and operations • Stickiness of knowledge transfer • Clashes between the development and operations cultures • Slower development • Ineffective operations • Increased outages • More customer outages
Informality Excess	Excessive informality of procedures due to the relative absence of information management and control protocols	Unknown	• Low accountability of both development and operations staff • Information spillover
Unlearning	A new technological or organizational advancement or best practice (e.g., as part of training courses) that becomes infeasible when shared with members unwilling to change	Experience diversity	• Lack of engagement • Gradual loss of knowledge or best practices
Organizational Silo	Siloed areas of the developer community that do not communicate, except through one or two of their respective members	High decoupling between tasks Lack of communication	• Decaying communication • Degradation of socio-technical congruence • Tunnel vision • Egotistical behavior

(*continued*)

Table 10.1 (continued)

Community smells (adapted from [Tamburri 2016] and [Tamburri 2019])

Smell	Description	Cause	Consequences
Black Cloud	Information overload due to lack of structured communications or cooperation governance	Lack of boundary-spanners Lack of sharing initiatives	• Misinformation • Obfuscated project vision • Lowering of trust
Lone Wolf	Unsanctioned or defiant contributors who carry out their work with little consideration of their peers, their decisions, and communication	Subgroup homophily	• Developer free-riding • Unsanctioned architectural decisions • Poor maintainability • Frustrated team members

automotive, mobile phones, information systems consulting, healthcare informatics, banking information systems, food production, and electronics), all of which claimed to be practicing agile software development, community smells were found in *every* project. The most frequently occurring smells were time-warp, cognitive distance and DevOps clash (which occurred in *every* team surveyed). These community smells can incur technical debt in your project:

- **Time warp** leads project members to assume that communication will take less time and conscious effort than it actually does, resulting in poorly communicated decisions and coordination problems.
- **Cognitive distance:** requirements are not communicated clearly, leading to the delivery of a buggy project that will be patched afterwards rather than built correctly the first time. The most senior engineers might not help the junior ones, leading to suboptimal code (e.g., more complex algorithms affecting system performance).
- **DevOps clash:** clashes between developers and operations personnel leading to slower development process (e.g., Git repository is very slow), slow (or impossible) to deploy products.

Furthermore, in this study there was a strong correlation found between community smells and self-reported architecture smells, which shows that social debt and technical debt are tightly related. In one study, this correlation was found in 80% of the projects surveyed. Finally, and perhaps most depressingly, the number of smells increased linearly with project age. That is to say that projects get worse, not better, over time and with additional experience. Hence, this is a phenomenon that should be identified and managed—it is not going to get better on its own. Like technical debt, social debt can be managed and we need to enact a conscious plan of action to manage and reduce social debt.

10.3 Managing Social Debt

Our goal in software development is not to avoid debt but to manage it. Sometimes debt is inevitable, necessary, or even desirable (in the near term). To be able to effectively manage social debt we need to align the project's software architecture to its team structure (or the other way around). Anything else violates Conway's law, and leads to suboptimal outcomes. Thus an architect or project manager should be thinking about and monitoring the social debt—and hence the sociotechnical congruence of a project.

The major activities of the architect or project manager to manage social debt need to center around:

1. Building a community
2. Making the community function
3. Continuously tracking and fostering the community

In terms of *building* a community, managers need to strive to make their teams *contiguous*. In an office where everyone is co-located, this is relatively straightforward—place the teams in close proximity to each other and ensure that they have the right incentives and opportunities to interact. But in distributed development or in cases where large numbers of developers telecommute, the luxury of physical co-location may not be possible. In such cases the teams need to be *electronically* contiguous, connected by shared repositories, mailing lists, video chat, instant messaging, virtual standup meetings, and so forth. In fact, electronic communities of practice have been shown to outperform physical ones, so this is probably a good practice even for physically co-located teams. Today, many companies (such as GitLab or Stripe) are moving towards a model where team members are located remotely, which also shows that remote teams can perform as well as local ones.

But how do you ensure that the community actually functions? We discuss some specific steps in chapter 11. First and foremost the architect, tech lead, and/or project/team manager should be active members of the community—coaches, mentors, advisors, and of course leaders. There is no substitute for leading by example. If you want your developers to communicate clearly, honestly, openly, and frequently, you must be seen as doing this yourself!

Even if you are providing a sterling example of how to be a good project citizen, problems may arise. The bigger the project, the higher the probability that problems will arise and that you may not know about them, or you may not know about them until they have reached crisis proportions. So how do you measure and monitor social debt (and manage it) in real time?

One possibility is to use a tool such as CodeFace or CodeFace4Smells[1] to track how the community is growing and how the culture is faring with respect to the community's proclaimed objectives. CodeFace and CodeFace4Smells are both free and open-source. These tools capture data from a number of project data sources, such as revision control systems, bug-tracking systems, and mailing lists. They then analyze the captured data to provide insight into your project's social networks and social activity. For example, CodeFace can identify the most central contributors, the most active discussion topics, and the prominent communication structures, among other things. Building upon this information CodeFace4Smells can identify (as its name suggests) community antipatterns: the social smells. The advantage of using tools such as these is that they can be built into an organization's standard pipeline. For example, this analysis could be run nightly or weekly, producing a report for project management. Having this information produced regularly in an on-demand or periodic basis gives managers and architects the information that they need to monitor and manage social debt and to react to emerging trends before they become major project risks.

Very often, such tools are not necessary, and it is possible to start detecting community smells by having private conversations with project stakeholders. For example, regular interviews between management and contributors (such as regular 1:1 weekly or biweekly meetings) are a great way to discuss potential issues among developers, their teams, their sister teams, or their management. These meetings should include questions about social debt and social smells to determine if there are problems with communications, problems with decisions and decision-making, problems with knowledge-hoarding, problems with bad behavior of others, and so forth. This is really for managers to show that they welcome any feedback, as developers often fear speaking up about problems and they avoid difficult discussions about organizational and communication issues.

Although there are, as yet, no precise quantitative estimates of how much damage can be connected to community smells, it is easy to see that they could have serious consequences (see, for example, Box 10.1 for a war story of how social debt hamstrung a project). These social and organizational phenomena can spread quickly throughout your network. For example, if a recurrent delay of two hours is connected to a certain community smell (and delays of two to four hours are not uncommon) and this community smell affects at least six different people in an organizational network, then that delay may end up affecting the entire friend-of-a-friend network!

Box 10.1
Voice of the Practitioner: Julien Danjou

> A large number of the maintainers quit the team over the years and left the remaining maintainers with the burden of supporting the code base. When none of the maintainers have interest nor knowledge in a feature, the code can actually become a burden. A large amount of development cycle was spent resolving that technical debt. Marginal features were deprecated and then removed; sometimes removing code is actually the easiest way to shrink your debt. On the other hand, widely used functionalities were taken care of, and issues were resolved by the maintainers. The fact that a development team has to spend a large amount of its time deprecating features, cleaning code, and upgrading API has a wide impact on the number of new features that it can deliver. There's always compromises to do; in our case, we often picked technical debt resolution over new code due to the shrinking size of the team. That might pose a problem as it can also prevent the project from getting interest from new contributors. Yet, those new contributors you don't have could help you maintain the project. Tough calls.
>
> —JD[2]

10.4 Avoiding Social Debt

Like technical or financial debt, social debt, if left untreated, will grow over time. For example, in one recent study by Damian Tamburri and colleagues, it was noted that the number of community smells increases linearly with the number of months that the teams work with agile methods. We have no reason to believe that agile methods are unique in this respect.

As stated above, to manage social debt you need to monitor your community, particularly if you have many external dependencies (such as dependencies on open-source products) or if your product is growing. As you grow you will typically include more products, which means more dependencies, often on other teams that you do not control. Growth makes the need to monitor social debt even more pressing. In such a case your organizational structure now depends on the organizational structures of communities that you do not control.

For example, when the Apache Mahout project decided to move away from Hadoop (and Map/Reduce) it was, at least in part, due to their perception that the Hadoop community was not able to maintain quality guarantees. When Joomla split off from Mambo it was over disagreements regarding project qualities and direction, and over who had control over what project aspects. Essentially the project had fractured—particularly between management and development—and the splits proved to be irreconcilable.

Monitoring a project—perhaps by tracking patterns of behavior in your project social networks or by anonymous mood-polling—may give you insights and tell you that something is going awry, but it does not tell you *how* to manage. We have seen success (and failures) in both highly formal and highly informal organizations. Fortunately there is some guidance in terms of project factors that have a consistent correlation to quality—particularly architecture quality. For example, the cohesion of a software community appears to be consistently important. A cohesive team is one with tight collaboration and communication. Having a highly cohesive organization results in a lower occurrence of architecture issues. This may be explained by the increased flow of architecture knowledge that naturally occurs in a tighter, more cohesive organization.

Fortunately there are strategies for managers and architects to address and reduce the consequences of community smells, although these fixes are often highly context-dependent. Table 10.2 maps some of these smells onto possible mitigations.

For example, if you are afflicted by the time warp smell, you can put into place better coordination and communication of architecture decisions. You can ensure that your lead architect is actively coaching team members, understands the development challenges, and is not just an ivory tower architect. You can also devote more resources to risk engineering.

If you detect the cognitive distance smell you can insert professional communication intermediaries into your team and process. You could also institute a practice of buddy pairing—pairing more experienced team members with less experienced ones. You can stimulate knowledge exchange via workshops, presentations, and living documentation in, for example, project wikis. If you detect the newbie free-riding smell you can, again, engage your lead architect as a coach or set up anonymous mood-polling so that you can detect such suboptimal behaviors early and with less emotional cost on the part of employees, particularly the newer ones.

Table 10.2
Community smell mitigations

Smell	Cause	Possible Mitigation
Cookbook Development	Thinking in an old framework—e.g., the waterfall model.	• Training and developer education programs; sponsored e-learning classes • Invited lectures to introduce developers and managers to new techniques • Personal project time to incentivize developers to learn new techniques, frameworks, and methods
Time Warp	A change in organizational structure and process that leads people to wrongly assume that communication will take less time and that explicit additional coordination isn't needed	• Improve architecture decision communication • Improve coordination of decisions • Have a clear map of communication flow in the organization (who talks to whom and takes decisions), e.g., making its transactive memory system explicit
Cognitive Distance	The distance developers perceive on the physical, technical, social, and cultural levels regarding peers with considerable background differences	• Use professional communication intermediaries • Have social time (e.g., lunch, weekly meeting, offsite) • Regular cross-team presentation on what they are working on • Hold workshops and presentations • Living documents • Outsource ontology alignment efforts to data governance companies (e.g., Collibra Inc.[3])
Newbie Free-Riding	Newcomers being left to themselves regarding understanding what to do and for whom, with the consequent free-riding of older employees (that is, the economic free-rider problem applied to software engineering)	• Coaching • Regular 1:1 interviews or daily stand-up meetings • Mood polling • Review retrospectives (e.g., looking to see which code submissions get many comments)

Power Distance	Lack of architecture knowledge sharing	• Give important tasks to junior developers and ask senior developers to mentor them • Incentivize senior developers to have empathy for junior ones, and help them to onboard and get more knowledge about the system
Disengagement	Lack of engagement in development Lack of curiosity	• Ask any tester (e.g., QA department) to break the product and demonstrate that part of the product was not complete • Allow engineers to focus 10% of their time to address technical debt and improve the overall code • Create more diverse teams (e.g., through gender and culture balance)
Priggish Members	Subgroup homophily Lack of engagement in development	• Show priggish members the productivity wasted addressing issues with little consequence • Ask team leads to set a bar where requests from priggish members are challenged during review, lessening the incentive to be priggish
Institutional Isomorphism	Excessive conformity to standards Lack of innovation Using formal structures to achieve community goals Rigid thinking	• Challenge processes and/or structure • Rapid prototyping of a new idea to see if it improves existing issues (or creates more) • Get data from other organizations and check if your decisions make sense.
Hypercommunity	Mergers and acquisitions Corporate restructuring	• Restructure and delimit roles and interactions, e.g., through NoOps • Abandon/avoid some communication channels

(*continued*)

Table 10.2 (continued)
Community smell mitigations.

Smell	Cause	Possible Mitigation
DevOps Clash	Distance among sub-teams, e.g., geographic, cultural, power, etc.	• Give ownership of DevOps to one entity (group/person) • Unify the code base (especially including infrastructure code) • Consider mitigations for power distance and cognitive distance
Informality Excess	Inexperience Lack of contribution policy and standards of behavior Missing or improperly disseminated architectural knowledge	• Define processes and incentivize developers to use them • Audit and log developer access and activity, including the ones deviating from organization processes • Block access to any user not following organization guidelines
Unlearning	Experience diversity	• Identify champions and evangelists for technologies and processes who are willing to lead and train others • Convince members with metrics about the efficiency of the new technology or process
Organizational Silo	High decoupling between tasks Lack of communication	• Organize periodic group meetings between silos • Have regular meetings between group leads
Black Cloud	Lack of boundary-spanners Lack of sharing initiatives	• Communicate information via protocols that reach appropriate members of the organization • Get feedback from members about protocols
Lone Wolf	Subgroup homophily	• Identify and sanction lone-wolf behavior • Pair lone wolves with other members

10.5 Summary

Social debt is real and it affects all projects, to greater or lesser extents. Various processes for managing software projects, such as iterative, agile, or waterfall, all succumb to social debt, because this form of debt is inevitable; it is inherent in the tribal ways that we humans interact. Older projects are, not surprisingly, more affected by community smells than younger ones.

But we do not need to simply accept that debt happens. Fortunately, you can automatically detect many community smells that identify specific kinds of social debt. There are mitigations that you can put into place to manage many of these smells or even to avoid them completely.

Notes

1. https://github.com/siemens/codeface; https://github.com/maelstromdat/CodeFace4Smells.

2. The full text of this and other Voice of the Practitioner sections can be found in the Appendix under the relevant name.

3. https://www.collibra.com/.

Further Reading

The original version of Conway's law appeared in over fifty years ago: M. E. Conway, "How Do Committees Invent," *Datamation* 14, no. 4 (1968): 28–31. His saying can be found on page 31. Other important work on the human aspects of software development include Tom DeMarco and Timothy Lister, *Peopleware* (Addison Wesley, 2013); Gerarld Weinberg, *The Psychology of Computer Programming* (Weinberg & Weinberg, (2011); and perhaps best known, Fred Brooks Jr.'s book describing the project management problems IBM encountered building the OS/360 system, *The Mythical Man-Month* (Addison Wesley, 1995).

There are many examples of how projects can fail due to nontechnical problems—problems centering on business, management, and social issues. One such example can be found in: "'*The Canary in the Coal Mine . . .* ': A Cautionary Tale from the Decline of SourceForge," *Software Practice and Experience*, July 14, 2020.

For decades, much has been written about the role of networks and communities of practice. Some early influential work includes M. Tushman, (1977). "Special Boundary Roles in the Innovation Process," *Administrative Science Quarterly* 22 (1977): 587–605; and R. Teigland and M. Wasko, "Extending Richness with Reach: Participation and Knowledge Exchange in Electronic Networks of Practice," in *Knowledge Networks: Innovation Through Communities of Practice*, ed. P. Hildreth, C. Kimble (IGI

Global, 2003), chapter 19. Also, the work of E. Vaast has shown that electronic communities of practice can outperform physical ones: "O Brother, Where are Thou? From Communities to Networks of Practice Through Intranet Use," *Management Communication Quarterly* 18, no. 5 (2004).

The community smells presented in this chapter first appeared in D. Tamburri, R. Kazman, and H. Fahimi, "The Architect's Role in Community Shepherding," *IEEE Software* 33, no. 6 (November/December 2016); and D. Tamburri, F. Palomba, R. and Kazman, "Exploring Community Smells in Open-Source: An Automated Approach," *IEEE Transactions on Software Engineering*, 2019. Some of the mitigation strategies discussed in this chapter were also presented in those papers. A study showing how prevalent those smells are in real-world projects is: D. Tamburri, R. Kazman, and H. Fahimi, "Organisational Structure Patterns in Agile Teams: An Industrial Empirical Study," arXiv:2004.07509.

A recent survey investigating the occurrence of community smells in agile teams and their correlation with architecture smells can be found in: D. Tamburri, R. Kazman, and W. van den Heuvel "Splicing Community and Software Architecture Smells in Agile Teams: An Industrial Study," *Proceedings of the Hawaii International Conference on System Sciences (HICSS) 52*, Wailea, Maui, January 2019.

A paper where the costs connected to social debt were actually quantified in a case study can be found in: D. Tamburri, P. Kruchten, P. Lago, and H. van Vliet, "Social Debt in Software Engineering: Insights from Industry," *Journal of Internet Services and Applications* 6, nos. 10:1–10:17 (2015).

Finally, a study discussing strategies for managing social debt can be found in: D. Tamburri, "Software Architecture Social Debt: Managing the Incommunicability Factor," *IEEE Transactions on Computational Social Systems* 6, no. 1 (February 2019).

11 Making the Business Case

To business that we love we rise betime,
And go to't with delight.
—William Shakespeare, *Antony and Cleopatra*

This chapter discusses business-facing challenges with managing technical debt from a practitioner's perspective. We show how to help nontechnical people understand (a) what technical debt is, and (b) the importance (and practicality) of addressing it. We break the challenges into three categories: metrics for managers, concrete actions to take, and how to bridge the gap between the technical and the social.

As we said in chapter 1, the original use of the term "technical debt" was a metaphor that was employed to justify the need to continually refactor and rearchitect software implementations. Ward Cunningham's use of the term has focused on what was, and remains, a ubiquitous problem with software: organizations seldom dedicate sufficient time and resources to improving software after it has been released. Software is extremely malleable, and at the same time, the problems that we expect software to solve are often poorly understood. This lack of understanding causes our software to constantly change and evolve. As we make clear, technical debt is also vital for software projects to move quickly, and it is an inevitable part of software development. Thus balancing the cost and value of technical debt is essential. In most organizations this balance is the domain of business.

The malleability of software makes it easy for an organization to put something together, release it, and put it into production, without understanding exactly what was just released. Management has traditionally had a tendency to think of shipped software as completed. However, by the

1970s it was already apparent that this was a fallacy. Manny Lehman, Laszlo Belady, and Burton Swanson, among others, recognized how important it was to continually maintain software, lest it grow stale and lose its suitability for purpose. Then, as now, convincing those holding the purse strings that rework was necessary was challenging. This mentality—that shipped software needs no further work—is caused by two problems. One, software is often treated as an expense, not an investment. And second, those who manage software projects often come from domains where maintenance needs are much less frequent.

The first issue is that most executives treat an IT project, particularly an internal project, as an incurred expense, much like buying electricity or envelopes. Naturally this leads to a view that these expenses are to be minimized where possible, since the return on investment is minimal. This is known as the cost center view of software.

The second issue is that many managers have backgrounds in very different projects (e.g., managing capital projects in manufacturing). If we are to create a new warehouse, the expense of building the warehouse is a one-time cost, and depreciated using an existing, well-understood depreciation schedule. While the warehouse might need some maintenance eventually (such as replacing the roof), this maintenance is understood to be rare and in the distant future. Software, by contrast, requires ongoing investment to maintain design integrity and to adapt to changing knowledge about the problem. Finally, although for many companies software was not initially seen as a competitive advantage (it would run internal functions such as payroll or inventory), increasingly these companies are in fact using software to differentiate themselves. Software is a competitive necessity and even a source of competitive advantage. Incidentally, this is frequently cited as the reason that major software companies (such as Facebook or Amazon) are more effective: they are led by people who get software in their very hearts.

This chapter examines how we can align our technical understanding of technical debt with business values and actions. We begin by examining business-facing ways to identify the debt. The chapter then looks at what management can do to take a strategic approach to technical debt, and it concludes with two approaches to better align technology and business.

11.1 Identifying Technical Debt with Metrics

11.1.1 Quantify the Economic Impact of Technical Debt

One of the primary reasons for the success of the technical debt metaphor has been its ability to bridge the language gap between business and technical staff. Putting software design problems in financial terms has greatly aided in making the problems of code quality tangible to nontechnical stakeholders. The notion that software loses value if not maintained—that it depreciates—has been well-known to technical software people since the 1970s and 1980s (based on work by pioneers such as Laslo Belady or Barry Boehm). Yet today, in most companies, software as an asset is poorly understood, and the idea of depreciation—commonly used in other assets, such as plant machinery—is rarely used in software assets, other than off-the-shelf software like Microsoft Office. Depreciation should apply but rarely is.

It is possible to extend this metaphor too far. Figure 11.1 (a repeat of figure 2.1) shows how we can think about technical debt with respect to what management cares about. In most companies profits and losses are the primary concerns. Spending time working on refactoring is a real financial expense; it is money not spent working on new features or fixing bugs. In nonprofit organizations, the real cost is in time. These costs are immediate and concrete, whereas the benefits are (usually) vague and long-term. As a result, management seldom approves rework to pay down technical debt (see box 11.1 on cost estimation and where we think it fits). We therefore crucially need to explain how technical debt can affect the ultimate value of our software system and consequently the company's bottom line. Part of this is explaining how this debt is an anchor weighing down productivity and innovation.

Figure 11.1 positions the notion of technical debt on a value curve. The y-axis captures the value of the delivered software and the x-axis the time dimension. Obviously our aim, as the software development group, is to maximize the value (since that will translate into higher pricing, more users, more clicks, etc.). Thus our task in conveying the importance of technical debt to management is to capture a single, simple value formulation. In this formulation, that implies measuring not only the value of functional and quality attribute properties of the code but also costs such as the cost of a delayed release, the cost of implementation and deployment, and finally, of most interest to us, the cost of rework. Rework is the cost required to repay the debt that has accumulated over time.

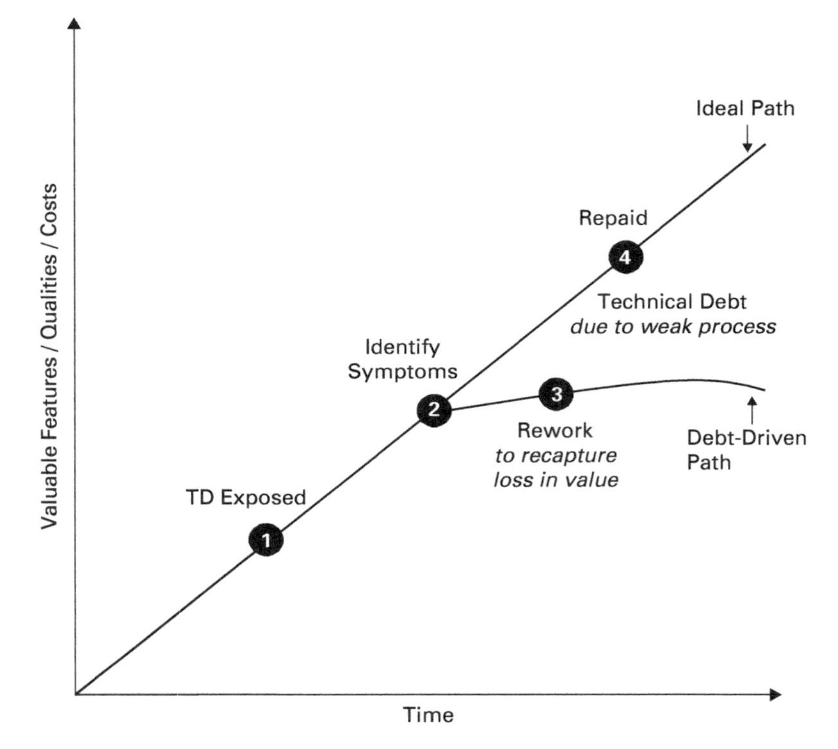

Figure 11.1
Technical debt and the software value curve.

We say that the debt is the amount of principal that must be repaid. For example, if we used a suboptimal routing algorithm in our code so that we could deliver that code more quickly (increasing technical debt), our principal is the cost to implement (and test, and release) a better version of the algorithm. Having a principal implies there is an associated interest cost. Interest in the technical debt world refers to the ongoing payments we have to make to carry the debt. The easiest way to think about interest is that it is the small, nonprincipal-related bug fixes, communication delays, misunderstandings, and slowdowns that we are responsible for. Interest costs are those issues that incur cost but provide no value. To return to the routing algorithm example, our interest costs in that case would be the performance hit that we take from the lack of optimality of the algorithm. It might also be seen in higher CPU loads, more data-center space, and so on. As with financial debt, the interest costs should go down as we repay the principal.

The simplest way to identify technical debt to management, then, is to think about the software system in this way: simply total the various costs and values (see box 11.1 on cost estimation for more details). Then report this to the management as the current value of their investment. Track the trajectory over time, as is shown in figure 11.1.

The curves in figure 11.1 diverge when the interest costs we pay on our debt begin to have a significant impact on the value of the product. Again, some of this may be acceptable if the other values are increasing (such as a reduced cost of delay). Not paying off the interest costs, though, and not repaying the debt, eventually lead us to a debt-ridden, negative trajectory. It is this divergence that is important to identify and report to management. Note that the area under the curve represents the total return on investment of the software effort.

The numbers in circles shown in figure 11.1 capture when debt is noticed, incurred, and repaid. At point 1, we incur the debt; perhaps the decision is made to release quicker rather than working on refactoring. At point 2, the symptoms of that choice are visible, for example, more bugs are reported. At point 3, we begin rework on the software to pay off the debt. If we do that successfully, the value of the software returns closer to the ideal trajectory (point 4).

While several dashboards exist to automate this reporting, we suggest that the most intuitive way to identify and track the debt is to simply make it part of your existing reporting processes. For example, this type of trajectory analysis can be automated, provides important feedback to a project manager, and would be useful input to an annual budgeting exercise.

11.1.2 Justifying Technical Debt Remediation

As many managers and executives consider software to be a utility and a cost center, it can be hard to make a case to invest in refactoring efforts to reduce technical debt, as it does not show any immediate value. Lowering technical debt is making a long-term investment. And, as we have said, the costs of this investment are immediate and concrete, whereas the benefits are long-term and uncertain. Humans, as a species, tend to pay far more attention to short-term needs and desires than long-term ones. That is why New Year's resolutions to lose weight and exercise more are easily

Box 11.1
Cost Estimation and Technical Debt

As long as we have been writing software, developers have been asked to estimate how long it will take. It's a natural question, but surpassingly difficult to do in the context of software development. The simplest mechanism is to get some experienced people together and to simply guesstimate how long or costly some task will be. A variation on this is to pool the guesses and throw out the high and low ones, averaging the remainder. These back-of-napkin estimates turn out to be surprisingly accurate.

In the Scrum process, estimation is done using planning poker. User stories are assigned points based on complexity, and the possible values are taken from the Fibonacci sequence (1, 1, 2, 3, 5, 8, 13).

Much more sophisticated estimation approaches are possible; Barry Boehm has pioneered much of the work in this area with his book *Software Engineering Economics* (Prentice Hall, 1981).

Finally, in response to the fact that many estimates are either wrong or out of date before they can be used, the #NoEstimates movement advocates for dispensing with estimates altogether. Their argument is that software's complexity is too great to ever estimate anything. Better, they argue, to get management to buy in to a truly agile and iterative process.

shed: because the couch looks comfy and the dessert tray looks yummy and those difficult choices can wait another day. Humans are *homo sapiens,* but definitely not *homo economicus*—rational decision makers—when it comes to economic decisions. However, we can take steps to improve the odds of making good decisions regarding technical debt.

The first way to achieve this, and to justify refactoring, is to create an estimate of the actual cost of the technical debt, answering the question: How much is our accumulated debt costing our project right now. Several tools, such as those described in chapter 4, report the cost of the technical debt principal (as defined in the introduction of this chapter).

People can track debt even without a tool: Are bug rates going up or down? Is the average time to fix a bug going up or down? Is our feature velocity (normalized by team size) going up or down? The choice of tool matters, since all coding issues do not have the same remediation cost (e.g., the cost to fix coding syntax is lower than upgrading a dependency). Thus making several estimates using several tools and approaches is vital to get a triangulated sense for the true cost. This financial figure maps the technical debt issue into a dollar value that is understood by management. Now take those costs and work backward using incident analysis or root-cause analysis, for example, to determine the actual source.

The second argument is related to efficiency. Codebases with significant technical debt will take more time to deliver a new feature. Most of the time, bug fixes or new features are delayed by such a codebase. It can not only expose the company to security issues (e.g., it can take weeks to migrate to a new library after a vulnerability is discovered) but also affect customer trust (e.g., features being repeatedly delayed for weeks). This ultimately has financial consequences (e.g., losing business). Executives need to decide whether to make the near-term investment to mitigate the long-term risk.

The third argument is related to human resources and retention. Poor code bases just make developers miserable. If code is hard to understand, modify, test, or build (or several of the above), developers will be frustrated and will not be effective. This will drive down the morale of the team. Ultimately, engineers will think they can make better use of their time by changing teams or companies. Teams working on a codebase full of technical debt will eventually become a nest of low-performing engineers with high turnover (with associated risks: loss of knowledge, decrease of velocity and agility, etc.). Investing in a better codebase is not a silver bullet, but it is a strong signal from upper management that they value developers and want to improve the situation for them and for their project.

As a rule of thumb, companies should spend 15%–20% of their time to remediate technical debt; this is what we have observed in healthy projects. Two of our interviewees said it was between 20%–25%; a Stripe developer survey (Stripe, 2018) showed about 30% of a developer's week was spent fixing technical debt, but this was seen as too high.

If your company does not invest in remediation, it is probably accruing debt fast, and bad times await. On the one hand, management should give time to their engineers to focus on debt, to stay efficient and agile. On the other hand, technical leads and managers must prioritize their technical debt and focus on the most important issues. It is important to reduce your technical debt, but it is totally unrealistic to expect that you are going to remove it all.

11.2 Management Actions

There are several concrete actions that we recommend for management to focus on to reduce and avoid technical debt.

11.2.1 Build a High-Functioning Team

As we discussed in chapter 9, social debt is critically important to understand, a point underlined by our Voice of the Practitioner, below. In our experience, and in books like *The Psychology of Computer Programming* by Gerald Weinberg (Weinberg, 1971), the majority of problems in software engineering can be traced back to human and social problems. To avoid incurring unwanted technical debt, efficient management will work to remove impediments to well-organized and efficient teams. This is not simply to set up an org chart that mimics that of better-known tech leaders like Spotify (and their squad/tribes model, which is a form of matrix management with technical and HR axes). It is rather to focus continually on community smells and the potential for social debt. One can track the team's health and efficiency with automated indicators (e.g., trailing indicators like commits per day or lines of code per day); but as with any metric, it is too easy to fall victim to Goodhart's law, where teams optimize the metric rather than overall value. For example, the number of commits might not show significant value if all of them are just configuration changes or minor improvements to code style or comments. The number of lines of code added or verified also depends on the language itself—some languages are far more verbose than others—or the style of the programmer.

Instead, we suggest that you use more qualitative approaches to check the overall health of the team, such as:

1. **Informal surveys and interviews:** take the temperature, do a mood-poll, ask your team how they feel and what their biggest challenges are. Make this survey anonymous so that people can openly share their feedback without fearing retaliation.
2. **1–1 with manager:** managers should periodically meet with the developers they manage and try to see if there is any pattern leading to inefficiency in the team (or more widely, in the organization).
3. **Intra- and interteam communication:** managers should look for communication within and outside the team and look for any signs of friction

or inefficiency. There are several communication channels that you can analyze to get good insights about friction points and other issues (e.g., code reviews and issue trackers are the main ones but mailing lists, Slack channels, or forms of group communication are also commonplace).

4. **Standups and retrospectives:** as part of the software process, short (<30 minutes) daily or weekly standups are a quick way to get a sense for technical blockers and obstacles. Retrospectives and postmortems are a form of reflection on how the past delivery cycle went.

Integrate the knowledge of a diverse set of team members who can provide multiple views on the customer problem and needs (e.g., marketing, sales, UI, backend, operations). Talking to only business and marketing, or only engineering, results in a flawed perspective that will ultimately result in producing a product that does not satisfy customer needs, is impossible to build and maintain, and that needs constant rework. Cross-team communication should happen early and continue through the lifecycle of a product, even after delivery.

More than any single metric (such as builds per day), effective management should focus on flow efficiency—overall feature cycle time, from identified feature to released software—rather than particular milestones in the lifecycle. Don't get distracted by short-term fixes. Prioritizing the time it takes to compile the code might only be useful if compilation is the actual bottleneck. Better to invest in resolving the bottleneck—perhaps the test time—instead. Lean approaches, as we discuss below, will reveal these bottlenecks as places where the work in progress drastically slows.

11.2.2 Focus on the Value Stream

In our chapter on requirements (chapter 3), we pointed out the importance of tracking features from requirements to release. The concept of a *value stream,* derived from lean manufacturing, reinforces this focus. Effective management will always be aware of where tasks are in the development cycle and what their value is projected to be. Effective management will also work to remove waste from the existing process. In the requirements context, this is most visible by the decision regarding which requirements to work on and which to ignore. We suggest the following concrete actions, based on our advice from earlier chapters.

1. **Track new technologies:** Effective technical leadership should be continually scanning the horizon of emerging technologies to identify

candidates that might provide a dramatic improvement in efficiency or quality. There is no silver bullet, but there are engineering improvements, and architects, technical leads, or PMs have a responsibility to educate themselves on these new technologies. Managers should understand the value of staying up to date and should support engineers to build expertise. For example, the shifts to DevOps, to radically decoupled software in the form of microservices, or even earlier, the shift to object-oriented programming, all gave substantial benefits to the organizations that adopted them properly.

2. **Analyze DSMs to track possible remediation costs and benefits:** As part of design analysis, design structure matrices (DSMs) provide a view into the current maintainability of the architecture of systems. Is the design becoming cumbersome? Is it weighing down the project, increasing bugs, and reducing velocity? Chapter 4 discusses this in detail.
3. **Conduct code quality analysis:** Management should always have a high-level view of the current state of the software. Less important than particular metrics is the trajectory of those metrics and their relationships to competitors. Chapter 5 explains this in detail. See also box 11.2.
4. **Automate testing and support coverage tracking:** Test automation is important, but it is unhelpful without confidence. A good suite of useful, information-rich tests is important to give evidence as to the state of the software (as discussed in chapter 6).
5. **Embrace a DevOps culture:** Once the tests exist, shifting to automated deployment environments dramatically reduces cycle time. But for this to work you must ensure that your test environments mimic your production setup. This will make deployment, and eventually delivery to customers, more seamless and trouble-free. Ensure that all code flows through this continuous integration (CI) process (see chapter 7).
6. **Strive for observability and deployment metrics:** Measure the value of your CI approach by ensuring each deployment is tied to measurable outcomes. A/B testing or blue/green testing are two ways to ensure that the value of a given feature is easy to understand. For example, you can track questions such as: Are more people using the feature? Have sales increased? Do users spend more time on site?
7. **Write docs and code together:** Keep in mind the value proposition for documentation (only write docs if using the docs will be more valuable

than the cost of writing them). Ensure that documentation and software evolve in parallel. Writing documentation at the time a decision is made is far easier than coming back and trying to remember to do it. Where possible, shift documentation closer to the code. That might mean simple techniques such as choosing good names for program elements, but it also means putting documentation into text-friendly formats (such as markdown) and storing the docs close to the code they document, as discussed in chapter 8.

8. **Track documentation usability:** As part of the documentation value proposition, conduct periodic sanity checks to verify that the documentation is still useful to the intended audience. Does the documentation suit the purposes to which it is being applied? For example, if the docs should help onboard new developers, are they actually being used for this purpose? Or do the newbies continue to pester the team leads? Chapter 8 covers this issue in detail.
9. **Understand the impact of machine learning:** These systems, discussed in chapter 9, bring with them new opportunities and challenges with technical debt. Effective management will be able to understand the risks, such as explainability, in these systems and use debt strategically to leverage machine learning as needed.

11.3 Aligning Technologists and Management

Consider an organization where management is only weakly aware of the technical challenges associated with shipping software and may have other backgrounds (e.g., finance), or other concerns (such as increasing sales).

Box 11.2

Voice of the Practitioner: Andriy Shapochka

> The most frequent impact caused by the technical debt is steep growth in the cost of project evolution combined with the low team development velocity. Technical debt also affects the number of defects existing in the codebase and hinders their troubleshooting and elimination. The side effect is also that people (especially senior-level engineers) are reluctant to join projects ridden with technical debt.
>
> —AS[1]

The first task for successfully communicating the need to deal with technical debt is to identify what the budgeting process for software investments is and how budget amounts are set. Is this an annual approach with fixed prices and fixed deliverables? Is it open-ended? Does management tolerate the inherently complex nature of software design and support iterative development? What are management's ultimate business goals? Answering these questions will make communicating with the business side smoother.

Our first task is to figure out how to identify and quantify the technical debt that we already have and to then communicate that news to management. We can use the techniques from other chapters to do this for our own needs, so here we focus on how to communicate that information to (presumably nontechnical) management staff.

11.3.1 Use Real Options

There has been extensive work on trying to assign monetary value to software. One such approach is the notion of real options for software, based on the work of Kevin Sullivan at the University of Virginia. An option in finance is the right to buy something. A real option is the right, but not the obligation, to take a business decision at some point in the future. Software modules can facilitate such decisions, and so they can be seen and analyzed as real options. Like everything in economics, an option has a financial value, and that value can fluctuate (my option to buy oil at $80 a barrel is worth much less if oil subsequently hits $70 a barrel). Options on stocks or cattle are relatively easy to price because we can make empirically derived assumptions about the underlying pricing model. This is very difficult with software, but the options framework still provides a useful frame for considering what software design, and the options embedded in a design, is all about.

As Sullivan writes, an options approach is to "view projects and products as portfolios of assets that can be designed to include valuable options, and then to manage such portfolios dynamically as uncertainties are resolved and new information is acquired over time." Managing technical debt in software development implies that team leads and product managers are able to understand their design and development options. For example, some designs may commit us too early to a particular technology. A fundamental idea in options trading is to defer the exercise of an option as long as possible (since the option has latent future value, as the underlying asset may go up more in value). In the case of software, premature commitment to a design may mean locking in to a framework that turns out to be hard to scale in a year.

Another way to think of this is to defer decisions until what Rebecca Wirfs-Brock calls the "last responsible moment." If, for example, the designer insulated the rest of the system from the knowledge of the framework by creating an abstraction layer, then it would be easy to change this framework later on. The creation of this abstraction is purchasing an option giving developers the right, but not the obligation, to exercise the option (i.e., to switch the framework) at some point in the future. Then the analysis question is this: Is the cost of developing the abstraction today (the cost of the option) more or less than the expected value of possibly exercising this option in the future?

Using an options approach is less about the underlying math and more about the model of thinking. You should focus on the idea that incurring debt today may be useful if it provides your team with a valuable option to take a new approach in the future. The greater the likelihood that you will switch and the greater the switching cost, the more likely you will be to purchase that option today.

Options also capture the notion that technical debt is nonlinear. For example, if part of the codebase is poorly written and hard to understand, but it never needs to be changed, then the real cost of this debt is zero. On the other hand, if that same debt exists in a critical part of the codebase, the cost can be astronomical. Chris Matts uses the notion of "unhedged call option" to reflect this nonlinearity. When we choose a particular design approach, there is typically no other alternative—we probably didn't implement more than one approach (hence it is unhedged). It is a *call* option because making this choice means we effectively sold our flexibility, and at any moment we may be asked to pay it off.

In theory, we can therefore rationally price out the risk that our option will be called, as well as the premium we gained by selling this option, and behave rationally—that is, only selling the option if the premium exceeds the risk. This metaphor captures the idea that today's shortcut might turn out to take longer or cost more than doing it the right way in the first place. This more closely reflects the uncertainty that pervades software development in complex and novel domains.

Ken Power, while working at Cisco as a software developer, introduced an approach like this to his team. Technical debt is explicitly tracked as a prioritization option, alongside feature development and quality debt (bugs).

The most concrete realization of this comes from Mik Kersten's book *Project to Product* (Kersten, 2018). Kersten identifies four types of work in software development: features, risks, technical debt, and bugs. Risks are things like security and privacy problems; technical debt is about refactoring and preparing for new features. Like at Cisco, the relative amount of each type of work is highly context-dependent, so there is no clear-cut criterion for how much of each to allocate but, as stated earlier, our experience suggests that 15%–20% of work should be on technical debt. Kersten mentions that if an organization is doing no work on any one of these, he would be concerned.

What is more important is the ability to track the progress over time. The innovation in Kersten's approach is that technical debt explicitly becomes a type of work that needs to be dealt with. Management could say that no technical debt work can be done in a given cycle, but that decision is explicit and (in principle) should be justified.

11.3.2 Track Software Design Moves

Design moves are another way to align the thinking of technical staff and management staff about technical debt. With design moves, each organization has a stock of design capital. Higher stocks of capital equal more options for problem solving. Technical debt reduces the ease with which these options can be chosen and acted upon. A *design move* is a choice, such as selecting a particular library for authentication management, or removing duplicate code, that either increases or decreases design capital.

The concept of design moves is a way to explain to product managers and business-facing staff why seemingly non-value-added work is vital. Developing new features is often of paramount importance to these members of the company. Requests to reduce technical debt by refactoring duplicated code into an external module are often met with skepticism that this effort will pay off. Design moves allow technical staff to explain that the work will increase the amount of design capital inherent in the project.

Narayanan Ramasubbu has done research into design moves in several company settings. In one example, he reports on how a company used a new configuration database to simplify product configuration, what he called an option-creating move. The practical impact of this move was to remove configuration related technical debt, making it quicker and easier to release new product configurations.

In a debt-creating move, he describes a company that moved to a multiplatform deployment setting. This increased adoption but incurred technical debt in support payments. Ramasubbu extended this idea into his framework for technical debt management. The essence is to first identify technical debt, then apply metrics to be able to evaluate how important and successful eventual remediations—the design moves—actually are. In a detailed longitudinal study, he explains how this approach worked in three large companies. Each company created risk registers, which are tools such as issue trackers that can make the risk visible. These registers are shared assets between business and technical staff. The companies then detailed mitigation plans, forward-looking roadmaps for reducing debt in the registers. Ramasubbu was able to show that this approach reduced software costs and increased value, even given the added cost to implement the registers and plan for the design moves. Implementing registers and tracking one's set of possible options had significant benefits for aligning the overall organizational approach to technical debt.

11.4 Summary

Communicating the nature of technical debt with the business-facing parts of your organization is vital in securing understanding and buy-in. It is important to understand the financial impacts, both positive and negative, that deliberate, prudent technical debt has. We have presented three mechanisms in which business can interact with technical staff in making decisions around technical debt:

1. Identify technical debt with metrics.
2. Support specific management actions in dealing with technical debt.
3. Use communication approaches that bridge the gap between technical and business staff.

While ultimately both parts of the organization have similar goals—profitability, mission success, and meeting external measures—language, culture, and knowledge frequently act as obstacles to recognizing this. Technical debt began as a natural metaphor for bridging this gap. Martin Fowler writes in his classic book *Refactoring* that when asked how to convince a manager that refactoring is important, he sometimes advises as follows:

> Many managers and customers don't have the technical awareness to know how code base health impacts productivity. In these cases I give my more controversial advice: *Don't tell!*

What he is suggesting, slightly tongue-in-cheek, is that dealing with code is what developers are paid to do. Therefore not every technical debt remediation necessarily needs financial approval, nor should it concern the business. You are a professional, and removing debt efficiently is something you should be doing as part of your daily responsible development practices. For larger rework phases, aligning your understanding of the value of technical debt mitigation with business people is essential.

Note

1. The full text of this and other Voice of the Practitioner sections can be found in the Appendix under the relevant name.

Further Reading

Ken Power's experiences are documented in "Understanding the Impact of Technical Debt on the Capacity and Velocity of Teams and Organizations: Viewing Team and Organization Capacity as a Portfolio of Real Options," in Managing Technical Debt (MTD), 2013 4th International Workshop, 28–31. IEEE, 2013.

For more on real options as a way to price out software design and construction, see K. Sullivan, P. Chalasani, and S. Jha, "Software Design as an Investment Activity: A Real Options Perspective," 215–262, in *Real Options and Business Strategy: Applications to Decision Making,* (Risk Books, 1999). There is also a lighter-weight read: Chris Matts's comic book on the topic: http://www.lulu.com/shop/chris-matts/real-options-at-agile-2009/ebook/product-17416200.html.

Narayan Ramasubbu is a business school professor at the University of Pittsburgh who conducts research into software development from a business perspective. See his paper "Integrating Technical Debt Management and Software Quality Management Processes: A Normative Framework and Field Tests," *IEEE Transactions on Software Engineering,* 2019, https://ieeexplore.ieee.org/abstract/document/8114229; and "Design Capital and Design Moves: The Logic of Digital Business Strategy," *Management Information Systems Quarterly*, 2013.

Lazlo Belady and Manny Lehman wrote an influential paper on software maintenance and evolution based on their experiences working on IBM's OS/360 project (the same project also spurred the writing of Fred Brooks's "Mythical Man Month" paper and later book): "A Model of Large Program Development," *IBM Systems Journal,* no. 3 (1976): 225–252. In this paper they propose the "laws of software evolution," many of which deal with the tendency of software to diverge from the intended behavior. For example, the "law of increasing entropy" states that "the entropy of a system (its unstructuredness) increases with time, unless specific work is executed to maintain or reduce it."

At the same time, Burton Swanson proposed "dimensions of software maintenance," namely perfective (refactoring design), corrective, and adaptive (accounting for changes in the external context). See E. B. Swanson, "The Dimensions of Maintenance," in Proceedings of the ACM/IEEE International Conference on Software Engineering, San Francisco, California, 1976, 492–497.

Lean software practices, adopted from the Toyota Production System, have been extensively catalogued. In particular, the writing of Mary Poppendieck: http://www.leanessays.com/2015/06/lean-software-development-history.html; Mary Poppendieck and Tom Poppendieck, *Lean Software Development* (Addison Wesley, 2003); Don Reinertsen, *Principles of Product Development Flow*, 2nd ed. (Celeritas, 2009); and David Anderson, *Kanban* (Blue Hole Press, 2010) have been instrumental in explaining how a philosophy developed at a car factory might be transferred to software.

Refactoring was made popular in a book by Martin Fowler, *Refactoring: Improving the Design of Existing Code* (Addison-Wesley, 1999), with a completely revised second edition released in 2018. The quote we use can be found on page 52.

Much has been written about Spotify's squad approach to teaming, which they have since greatly modified. The original write-up is at: https://blog.crisp.se/2012/11/14/henrikkniberg/scaling-agile-at-spotify. Mik Kersten's book "Project to Product: How to Survive and Thrive in the Age of Digital Disruption with the Flow Framework" (IT Revolution Press, 2018) summarizes Mike's practical experience with lean software development.

The topic of developer productivity and satisfaction has been a long-time concern. The first books on this subject were Gerald Weinberg, *Psychology of Computer Programming* (Dorset, 1971); and later, Tom DeMarco and Timothy Lister, *Peopleware* (Addison Wesley, 1987). Academic researchers currently directly study developer satisfaction and productivity. Graviton did a large survey that showed 219 factors causing unhappiness: Daniel Graviton et al., "On the Unhappiness of Developers," EASE, 2018, https://arxiv.org/abs/1703.04993). There is also emerging work on brain activity during programming and design tasks, for example Peete et al., "A Look into Programmers Heads," *IEEE Transactions on Software Engineering*, 2018.

Much has been written about behavioral economics and how humans are typically poor at making economic decisions in a rational way. Three excellent recent books on this topic are: Dan Arielle, *Predictably Irrational* (Harper, 2009); Richard Thales and Cass Sun stein, *Nudge: Improving Decisions about Health, Wealth, and Happiness* (Yale University Press, 2008); and Daniel Hahnemann, *Thinking Fast and Slow* (Farrar, Straus and Giroux, 2013).

Reading social media or consulting company reports such as the Thought works Radar facilitates understanding new approaches and technology options. Stripe's 2018 report, "The Developer Coefficient: Software Engineering Efficiency and Its $3 Trillion Impact on Global GDP," can be found at: https://stripe.com/files/reports/the-developer-coefficient.pdf.

Case Study D: Safety-Critical System

Summary and Key Insights

This case study looks at the specific problems technical debt generates in safety critical settings. Since the primary driving quality requirement is safety, technical debt takes on a new relevance for this system. These systems often use specific language, compiler, and design choices. Furthermore, safety critical systems are often (as in this case) also embedded systems, which make testing and deployment big challenges. We report on how this example system faced challenges with coding, testing, and deployment, and then show how the system might reconcile these challenges: using more modern tools, improving the test infrastructure, and working on documentation. These efforts are all necessary to reduce the ongoing technical debt the system suffers.

This (anonymous) system case study was based on personal experiences of one of the authors of this book.

Background

This case study is about a safety-critical system, which is one used on military, avionics, or nuclear systems, and where life or millions of dollars can be at stake. The particular software we are considering was written in C and performs critical operations for a satellite-based positioning system. Failure of this software might impact boats, planes, or trucks (failure to get location and directions, blocking hundreds of vehicles).

System Overview

Initial Design

The project started as a governmental project in the 1980s. Specifications were written in a text document and requirements were elicited using a spreadsheet. As usual for critical systems developed under government contracts, multiple independent contractors were responsible for different parts of the system, especially for testing (so that implementation and testing code is written by different developers, removing the scenario where testers only test success cases and write tests according to their understanding of the implementation, not the specification). Because of the large numbers of distinct organizations involved, social debt (see chapter 10) was inevitable.

Design, testing, and acceptance documents were all lengthy text or spreadsheet documents, as required by certification authorities. Unfortunately, these documents are as easy to understand as Egyptian hieroglyphs. As a result, it would not be surprising to have engineers that approved the software without having a clear understanding of the system but validated it because of time constraints or financial pressure. Engineers that are currently working on this project have a very difficult time understanding these documents as they need to know the entire context from the beginning of the project to fully understand the documentation.

Implementation

The system was designed to work on a specific processor family (PowerPC architecture). The system had to be deterministic and operate under strict, real-time timing constraints. At the time of the implementation, few languages were able to meet these constraints: mostly assembly, C, and Ada. The decision was to use the C language (the only alternative was Ada, as an assembly language would have been too complicated for such a project).

At the time of the implementation, it was standard to favor performance over readability. Also, software engineering methods and metrics were still emerging and there were no strict, formalized ways to write clean and good code.

The code was hard to understand for several reasons. The software was written with a custom C compiler for the chosen architecture. For the sake of performance (and because of the lack of optimization from the compiler), many coding tricks were used (such as intensive use of global variables and lots of duplicated code to avoid loops and function calls). As this software was written by different developers, the coding style was inconsistent and therefore, the software was hard to understand.

The project was successfully accepted and delivered. Tests were written, executed, and checked. These were end-to-end tests that required an engineer to compile, run, and observe the result manually. This was a manual and intensive activity that could not be automated.

Justified Technical Debt

The project introduced what would be seen today as technical debt, especially on the coding and testing side.

At first, for optimization and performance, developers made intensive use of global variables. That choice made sense at that time, where code was focused on optimization while maintenance and readability were less important. There was also a heavy use of C macros in most of the code, which was a decision made to avoid function calls. These hacks were made for good reasons at the time: the system, being safety-critical, needed to meet its hard real-time deadlines and so every performance optimization was considered.

Because of the diversity of compilers (and probably the lack of a reference compiler for multiple platforms) at that time, the whole build system relied on a proprietary PowerPC compiler. The compiler ran on a particular class of systems (at that time, there were a more diverse set of operating systems than today) and targeted a specific set of processors.

As mentioned, tests were mostly manual: to check the software correctness, engineers needed to compile and run the tests themselves. A special spreadsheet document, required by the validation authorities, outlined the testing scenario: what test to compile/run, what values to test, and so on.

Maintenance and Corrective Issues

Over the years, bugs were reported, fixes were requested and applied. As the years went on, changes were more difficult to make because of the following issues:

- **Coding:** the use of shortcuts to optimize performance and execution time made changes more complex. For example, the use of global variables made it very hard to track the control flow of the software (e.g., what function changes what variable with what value). This increased the investigation time to find a bug and ultimately fix it.

- **Reproducible builds:** as the compiler was proprietary software running on a specific platform, it became difficult to rebuild the software that was originally delivered. The original compiler was no longer supported and it did not run on conventional operating systems. Therefore, building a new version of the software was very time consuming and only a few people knew how to do it.
- **Training:** the code itself was very hard to understand (in part due to a lack of comments) and most documentation was very high-level, not going into the technical details. As the product continued to be supported for years, developers needed to be trained. The learning curve was steep and it took months for developers to be ready to make changes on the software.
- **Testing:** tests were manually done, which is labor intensive. Tests were not run automatically at each change and rather were executed at the end of a maintenance cycle, when software is about to be delivered. This increased time to deliver a fix.

Actual Problems

Real problems arose years after the initial implementation. Specifically, the system would stop operating suddenly without any signal indicating any malfunction. Developers who did not write the original systems had to hunt for the bug that caused this issue. Because of the project status (and its related technical debt), it took months to fix these issues. While trying to fix the bug, the system experienced other shutdowns.

Coding Debt

The code was initially written using old coding patterns (e.g., use of global variables or macros to avoid loops) that decrease code readability and maintainability. The use of these patterns were justified at some point but also increased the time required to read and understand the program. New engineers took more time to learn and be familiar with the code and start to be active contributors. These can be seen as increasing interest costs on the initial technical debt.

It also increased the time to make a modification consistently across the code base. As blocks of code were duplicated, maintainers had to track potentially duplicated code that also needed to be modified (which meant that someone had to distinguish which block of duplicated code was impacted by an issue).

Testing Debt

On top of the actual coding debt, there was also testing debt (see chapter 6): tests were started manually once code changes were done. There were no automated tests that ran consistently on all the components and checked their correct integration. At best, there were some unit tests available on some components, but that was definitely not enough to accept a change.

This debt was partly caused by the development process: as components were developed by independent contractors, one contractor would not have access to the component of the others (as the integration tests were often done by other independent contractors). This limited the testing opportunity.

Deployment Debt

Lastly, there was an important deployment debt (see chapter 7). First, the system could be built only on a specific platform, which reduced developer agility because the system cannot be built on the developer's own laptop. There was also no possibility to test the system locally, and each run required compiling the code and attempting to run it on the hardware. This made new releases very slow to deliver.

How to Address the Technical Debt?

Our goal is not to say that everything that was done was bad. As always, it strongly depends on the context. Original implementation decisions made sense at that time considering the context (at that time, performance optimizations were required). However, technologies changed and evolved, and most of these original choices would not be valid today.

The original code does not constitute technical debt per se. On the other hand, by not updating the code and bringing it up to actual coding, testing, and deployment standards, it started to be outdated and became a pile of technical debt.

In the next section, we discuss how this debt could be addressed.

Use a Modern Compiler

The first good move toward eliminating the technical debt would be to replace the compiler (or use a VM/emulator), and use an up to date compiler that works on commodity hardware (x86–64 architecture). That would at least let developers compile and run code on their laptops. This would then enable developers to work by increments, without having to work on a dedicated platform. Of course, it is not easy to change compilers quickly, and it is necessary to check that the new compiler not only builds the system but also passes all existing tests.

Develop a Shared Testing Environment

The second major benefit would come by having the ability to automate tests and let developers run tests on their own laptops—which requires the project to have a modern compiler! (We discussed testing debt in chapter 6). There will be two major challenges to overcome to enable this:

1. Write and automate unit tests
2. Design a mocked environment for end-to-end testing

The first aspect (write and automate unit tests) would help engineers to be more efficient and to iterate quicker. Engineers would be able to run tests on their own laptops. That would not give full confidence that the software is working correctly (some tests—especially time-sensitive ones—require the real hardware to run), but it could cover more than 90% of the tests. This would cover only unit tests. The second aspect would bring the ability to engineers to go further than unit testing and execute integration tests. The mocked environment would replicate the real execution environment and simulate its processors, sensors, and other devices. Many developers actually do this by providing a version of QEMU, a standard emulation tool, with extensions to simulate particular hardware. That would let each contractor run functional tests.

On a project of this size, it would take months to write such tests. However, that would not only help in the future but would also help to discover new bugs and make the software more robust. As with most technical debt management choices, this would be a multiple year effort, but would definitely pay off. It means that the system owner needs to continuously invest in the software maintenance over the years to adapt it with current practices but also make sure potential issues will be resolved quickly. In this case, having tests would have reduced the time to fix issues that surfaced years after normal operations.

Use of Static Analysis and Periodic Refactoring

The previous recommendations were about making the system up to date with current practices and increase development efficiency (reduce iteration time, facilitate development on commodity hardware, etc.). Another action that would reduce the system technical debt would be to use static analyzers and fix the biggest issues such as the use of global variables, uninitialized values, or simply the fact that a lot of code was copied across the codebase (these issues are discussed in chapter 5).

Static analyzers surface different dimensions of issues such as software defects, software complexity, or code duplicates. Using them periodically, identifying issues, and fixing them would increase the system quality and avoid potential bugs to surface down the line. In particular, using them can help developers and managers see if the overall quality of the system is improving and help them notice when new bugs are introduced.

Improve Documentation Approach

One area of improvement is related to documentation—and more specifically, how to make documentation more useful for developers. As for all safety-critical systems, most of the documentation required by certification authorities was generated using text processors or spreadsheets. However, this documentation is very large and often hard to understand for newcomers.

There were several aspects of the documentation that could have been more useful:

1. **A getting-started guide:** a step-by-step guide that explained how to build, deploy, and run the software. As pieces were implemented by different contractors a long time ago, there was not a single, up-to-date place that explained how to get started. Just running the software was taking days for a new developer.
2. **High-level documentation:** a short document (less than ten pages) that would have shown the overall software (organization into packages, functions, etc.) and system (showing deployment aspects) architecture. This would have just given a high-level view, but this alone would have helped understanding where to look when modifying a particular functionality.

3. **Code documentation:** a lot of code was left undocumented. Without high-level documentation and lacking code documentation, troubleshooting and investigation consisted of many wasted hours reading and poking at code.

In such projects where documentation is required by certification agencies, there is always a temptation to not document more than what is required. The problem is that the documentation required by these agencies is often useless for developers. Instead, following the guidelines we discussed in chapter 8, we advocate the creation of developer-friendly documentation, a lightweight set of documents that helps to quickly understand the organization of the software. Its aim is to start fast, and to give a great high-level view so that any new developers can be familiar with the code within a few hours or days (at most).

12 Conclusions

> In real life endings aren't always neat, whether they're happy endings, or whether they're sad endings.
>
> —Stephen King

We began this book introducing technical debt as a metaphor. However, our intention was and is to move beyond the metaphor. Metaphors are useful for sure. They help in communication and in consciousness-raising. But we are, or should aspire to be, engineers—and engineers take a scientific approach to decision making based on evidence and analysis.

To help move beyond the metaphor and to convey a disciplined approach to dealing with technical debt, we adopted a common structure in each of chapters 3–9: identify, manage, avoid. Each of those chapters addresses a major engineering activity or management concern in software development such as coding, design, deployment, or testing. In chapter 9 we discussed how technical debt is a concept that can be applied even to emerging areas of software development, including machine learning systems.

Each of those chapters provides a practical set of methods and metrics and tools to the developer and manager to scientifically reason about technical debt—to identify it, to manage it, and to avoid it. These methods, metrics, and tools are not just our opinions: they have been validated in the real world and are in use in real-world projects today. That is why we titled this book *Technical Debt in Practice*. We strongly believe that the experience of practice is, in the end, the only true measure of the value of an idea in software development.

Now it is up to you to take those tools and techniques and start to identify and work on your debt. Remember that everything we have talked about must be gauged and prioritized based on the return on investment for your project. Not every debt item is bad, and not every debt is worth repaying. Repaying debt costs something, and every hour or dollar you spend on retiring debt is an hour or dollar that you are *not* spending on fixing bugs and developing features (which is, in the end, what your customers pay for). Technical debt can be something of strategic value when it is incurred prudently.

In many ways employing the tools and techniques we have presented here is the easy part. Once you have identified true debts—debts that really should be paid off—you need to convince your other project stakeholders that it is a good idea to do the grunt work of paying it off. Developing features and putting out fires seems much more attractive than refactoring your code base. But in the end fixing your high-priority debt items is just the right thing to do, and fixing it will pay you back many times over in terms of better productivity and a saner technical environment.

Most organizations, PMs, and developers do not do this today. The practices of identifying, managing, and avoiding technical debt are not widespread in the software development world and are, in all likelihood, not the status quo in your organization. So you need to be a change agent in your organization: convince your organization to take action on the debt that you know about, find out what you don't know, and deal with it before it causes major problems.

We showed with the four case studies—case study A through case study D—that technical debt is real. However, in many cases something can be done—or has been done—about it. Let's briefly recap each case study and its outcomes.

In case study A we looked at a modern product-focused software company and how it deals with technical debt. The company—Brightsquid—develops secure communications products for the healthcare industry. In this study we showed how the tools and techniques from chapter 4 could help a company identify and strategically manage its technical debt. We used design metrics and a tool that pinpointed architectural design flaws to find trouble spots in the codebase and to develop a refactoring plan to "pay down" the debt. By collecting appropriate data and presenting this to management along with expected ROI data, the Brightsquid architects not only figured out precisely what to fix but also managed to convince management to approve refactoring sprints. They paid down their most serious debts and life was much better as a result. Bugs went way down and productivity went way up.

In case study B we discussed how a large and rapidly growing company—Twitter—faced serious challenges with the technical debt incurred by early design and implementation choices. In its early days, Twitter faced many technical challenges for scaling its services, and the service was often down. Today, Twitter is one of the most robust services in the world. They accomplished this transformation by recognizing the debt and paying it down via large-scale design changes, including a shift to microservices and new frameworks. In addition they dramatically increased test coverage—paying down their testing debt—and made use of a continuous integration (CI) mindset and CI tools.

In the ALMA (Atacama Large Millimetre Array) study (case study C) a highly complex and uncertain domain was presented—that of scientific software for a complex and long-lived project. We showed how the project team struggled with and, in the end, managed and paid down various dimensions of debt including requirements, code, design, and testing. They brought their requirements debt under control by instituting a formal requirements tool and change tracking process. The ALMA project cannot be said to have completely managed their code and design debt, although they are conscious of it and are working towards reining it in. The ALMA team addressed testing debt organizationally—by fighting for more resources and by creating a distinct integration testing team. Finally ALMA faced social debt challenges, as this project is long-lived, spanning many national organizations, and it had to deal with a shift in team structure between telescope development and telescope operation. To deal with this form of debt they have instituted cross-functional teams that can help improve collaboration.

Finally, in case study D we presented a study of a safety-critical system where hard real-time requirements needed to be met and, as a result of this singular focus, substantial technical debt was acquired along other dimensions. This debt was never paid off by the development organization, and they suffered greatly from that decision. We added our own editorial comments to the end of that study discussing how they *could* have addressed some of this debt via static analysis and periodic refactoring, an improved documentation approach, and the development of a modern automated testing approach.

As you can see from these case studies, an important takeaway message of this book is that technical debt is much more than merely implementation debt. Concerns over the accumulation of debt in design and implementation (chapters 4 and 5) are where the metaphor began, for sure. Frequently design and implementation consume an inordinate amount of effort on a software project. But since we first began researching technical debt-related issues, we have come to realize that debt has *many* dimensions; ignore these at your peril.

This is a really useful perspective to take. If we restrict ourselves to only worrying about shortcuts in our design process, or our code, we fail to consider the broad set of suboptimal approaches that we may take in other important areas of software development. For example, getting the requirements process well defined is vital in understanding what you should be building, as we showed in the ALMA telescope case study. While you should expect the requirements to change (or at least, your understanding of them to change!), it is important to get a reasonable idea of what they could be, before committing to code, particularly for a large, complex project.

By rushing to implementation without understanding the requirements to the best of your ability, you may be taking on debt. Of course, we understand that some requirements simply cannot be known in the early stages of system development—that is not the same as debt. Debt is a problem when you *could* have understood the requirements better and produced a better product but chose (perhaps subconsciously) to not do that in your rush toward delivering something. Similarly, a badly rushed deployment process may destroy all the hard work of the design and implementation phases (as we described in our discussion of Twitter), by creating a system whose deployment is needlessly complex or inflexible.

We want to emphasize that identifying, monitoring, and managing these various forms of debt is not a panacea. You can still make mistakes, and your project might still fail, but this is true of any complex human endeavor that involves uncertainty. The point we are making is that you can dramatically improve your chances of being successful by paying attention—by making conscious choices about when, where, and how much debt you want to acquire and when to pay it back, if ever.

The reality is that if you are building nontrivial, long-lived software, you will and in fact you should have debt. Even so, you should be devoting a serious effort—perhaps as much as 20% of total project effort—to tracking, managing, and mitigating this debt. To do less than this is a foolish risk.

The final takeaway from our book is an important one, and one that you have probably heard before: if you can't measure it, you can't manage it. Management gurus have been saying this for decades. We have seen how a lack of visibility into the accumulating problems of technical debt often doom projects. So technical debt begins with measurement, and this measurement allows you to identify the problem. If you make this measurement a regular practice then managing the debt and even avoiding it in the first place (or at least avoiding its nastiest consequences) becomes a natural and normal part of your development process. Once you begin to think this way, you will never be able to go back to your old software development practices. We promise!

Appendixes

Interview: Marco Bartolini

Introduction

Marco Bartolini is a software engineer working on the world's largest radio telescope. He has a computer science background and has worked in supporting the computational needs of radio astronomy, including with INAF, the Italian Astrophysics Foundation. He is also a certified Scaled Agile Framework practitioner, which is a methodology used with the Square Kilometer Array.

Summary and Key Insights

In this interview, Marco discusses his experience with technical debt while working on the Square Kilometer Array (SKA) software project. SKA is using Python for user-facing data processing scripts, with C++ used in data-intensive operations. One of the big challenges with SKA is that the architecture is still in the planning phase, and thus it is much earlier in the lifecycle than other case studies and interviews in this book. Another challenge is that SKA is multinational and governmental, and so it involves many moving pieces.

In particular, Marco emphasizes the necessity to test and use up-to-date, well-maintained libraries and to be proactive in the management of debt. For example, SKA is using a form of the risk register we discussed in chapter 11 to clearly identify libraries that will need to be refactored in future iterations. Marco is a firm believer that early identification of technical debt is vital to properly managing it.

Detailed Interview

Please introduce yourself, including a brief description of your background and experience in software engineering.

Hi, my name is Marco Bartolini and I am a software engineer currently working as Software Quality Engineer at the Square Kilometre Array (SKA) organization. My background is in computer science and I've spent most of my professional career developing software systems in the field of astronomy.

Can you tell us how do you define technical debt and how it applies to your project?

It is very hard to define technical debt in a consistent way that is well understood by developers as well as business people, and that does not lead to excuses and compromises about the quality of the software system under development. Ideally, in our current project we log as "technical debt" only what Martin Fowler classifies as "deliberate and prudent" in his technical debt quadrant. That is to say that technical debt should happen only when we consciously adopt solutions that do not conform to our quality standards.

We continuously stress the importance of building quality into the software, as we develop it, and we do not think of technical debt as a shortcut to skip over quality aspects of the development activity. For example, we would not accept a feature that is developed with no tests just because we logged technical debt in our system. But there are cases when a suboptimal solution is inevitable and necessary. What's important from my point of view is that we try and keep technical debt under control, maintaining a system where we can estimate what's the amount of technical debt we are dealing with, and using capacity allocation to prevent it from diverging to an uncontrollable amount. Allowing for refactoring as part of the normal development activity is also a key ingredient in maintaining the debt under control.

Someone in our community (Tim Jenness, from LSST telescope) suggested to adopt the term "escalating risks" as a replacement for "technical debt," as suggested by Jessica Kerr in this twitter thread https://twitter.com/jessitron/status/1123310331957145601.

Tell me about your project: what the project does, how many users, the types of users (individual vs. company), is it open-source or closed-source, any special/challenging requirements, and so on.

The Square Kilometre Array (SKA) project is an international effort to build the world's largest radio telescope, with eventually over a square kilometer (one million square meters) of collecting area. The scale of the SKA represents a huge leap forward in both engineering and research & development toward building and delivering a unique instrument, with the detailed design and preparation now well under way. As one of the largest scientific endeavors in history, the SKA will bring together a wealth of the world's finest scientists, engineers, and policymakers to bring the project to fruition.

The project as of July 2019 is approaching the conclusion of its design phase, and construction is expected to start in 2020. [Note: the telescope passed Construction approval in September 2020.]

In order to realize this one-of-a-kind telescope, a number of software systems needs to be architected and developed. These range from the control systems operating the telescope, to supercomputers that will be used to process scientific data, to the complex logics that manage the lifecycle of astronomical observations. The software development activity is carried on by a large collaboration of different actors, including a number of research institutes from the member countries and industrial partners. This highly distributed and dispersed model of development is a challenge on its own: communication, transparency, and ease of access to relevant information all play a key role. Adopting an open-source licensing model and promoting common code ownership is one of the fundamental steps in this direction for our project.

From a software engineering perspective our most challenging requirement is certainly represented by the performance needed to deal with the amount of data produced by the telescopes. We expect the telescopes to deliver an aggregated amount of about 15Tbits per second of astronomical data, and these need to be processed in real time so that a manageable size of scientific products can be stored. For this to happen we will need a 250Pflops supercomputer that participates in the astronomical observations, elaborating the data in real time, leading to store 300PB per year of scientific data.

Everyone interacting with this telescope can be considered as a "user," from my point of view, so not only scientists looking at astronomical data but also engineers building the telescope or doing maintenance, as well as telescope operators. It is easy to think that thousands of people will interact with the telescope in a way that is mediated by software.

Tell me about the history of the project: When did it start? Who maintained it a few years back and now? How has the project been architected and managed over time?

Like many large science projects, the SKA has had a long gestation period. In fact, the construction project hasn't even started, but there had already been an SKA History Conference. The project traces its roots back to the early 1990s and in September 1993 the International Union of Radio Science (URSI) established the Large Telescope Working Group to begin a worldwide effort to develop the scientific goals and technical specifications for a next generation radio observatory. There followed a number of investigatory phases, coordinated via MoUs between institutes from various countries in 1997, 2000, and 2005, before getting its first project office in 2008. A legal organization for the development of the design was formed as a UK company in 2011, and the detailed design was developed over the next decade.

The project, as of July 2019, is concluding this design phase. During this phase the telescope has been designed as an aggregate set of different elements, coordinated by a lightweight central organization. Among these elements, some are definitely software-heavy, such as the Science Data Processor (SDP) and the Telescope Manager (TM). For these, we concentrated the design effort on the development and delivery of a well-documented software architecture, validating some assumptions via prototyping when possible in terms of time and resources. We are now in a phase where we are integrating these architectures into a single system of systems, incrementally developing an evolutionary prototype that contributes to validate and verify the architecture designed in the previous phase. In doing so, we are gradually pivoting from a document-based, earned value, stage-gated set of processes to a code-based, value-flow-driven, lean-agile set of processes unified around the Scaled Agile Framework (https://www.scaledagileframework.com/).

Tell me about the language: What programming languages are used? How do they interact together?

We have a very diverse set of systems, but our main programming language is Python version 3. This is defined as part of our standards, where we predict that Python will be used to the maximum extent, in every possible occasion. This is related to the fact that the language has reached a significant level of maturity and its widespread adoption within the scientific community, especially in the world of astronomy. While Python is our main choice, we have a number of systems where performance is key, so C++ is used extensively along with GPU libraries and compilers. Web-based interfaces are mostly realized using JavaScript. The SCADA system development is based on a middleware framework called TANGO (https://www.tango-controls.org/), that is realized via a lightweight use of CORBA and it supports C++, Java, and Python as programming languages. This is indeed the glue of our system of systems. A number of data format and transmission protocols are used to exchange information between the processing-heavy components. FPGA firmware is also developed for ad-hoc processing boards, so VHDL is a first-class citizen in our languages ecosystem.

How is the system tested? Do you use unit testing? Integration testing? Code coverage? What has been used since the inception of the project, and what testing has been added later?

Our testing journey has just begun, and our testing policies and strategies are being defined. We are certainly starting with unit testing, and we promote the adoption of test-driven development as a standard development practice. We are automating most of our tests from the start, and we are continuously testing and integrating our software components. We target a model of continuous integration and continuous delivery for our whole system. We want to test and integrate the whole software system in a virtual test environment, so the development of hardware simulators and software mocks will play a key role in our testing strategy.

What would you describe as technical debt on your project? What are the biggest debt points, and how does the team approach and address them? Can you quantify the impact of these debts?

Technical debt can happen in many forms. One of the difficulties we face as a project is represented by the quality of some highly used scientific libraries. Some of these software products are state of the art in terms of functionality, but they are often the outcome of layers and layers of research work, often resulting in poor maintainability and testability of these systems. But science libraries of new generation, such as astropy,[1] really shine under the aspect of software quality as well, so our life is becoming easier. When we accept to use something knowing its limitations, we usually create one or more features in our backlog representing the work needed to refactor or rewrite the product according to our quality standards. These features follow the natural lifecycle of estimation and prioritization as any other feature of the system, but we have them in a dedicated bucket (an epic in our system), so that we always know what's the total amount of debt and we can decide to force some allocation if we see it is diverging.

The research of extreme performances in some areas can also lead to building up technical debt by adopting bleeding-edge technologies and software solutions, which often suffer from a short lifespan. We always bear in mind that our software system is expected to be maintained and updated for the lifetime of the telescope, that is at least fifty years. This often ends in evaluating the tradeoff between the maturity of a product and its immediate performance benefits. Maintaining an open approach and adopting a set-based design is vital in making decisions at the most profitable moment.

Tell us a story of how technical debt affected your project: What were the impacts and the outcomes?

For our first prototypes we adopted a specific software library for astronomy calculation related to pointing the telescopes: we really liked the semantics of the exposed interface and how cleanly the API was exposed to the developers. Unfortunately the underlying computational core was based on a well-written but now discontinued library. The end result is a conscious choice to adopt the library as-is, and in the meantime allocate some capacity to port it to a different computational core while maintaining the same API. In this particular instance this also contributes to improve the overall ecosystem of the astronomical software development, so we see it as a way to give back to the open-source community.

Looking back at your project, what you could have done better in terms of technical debt: prioritization, refactor, and so on?

One of the key lessons learned from the start of the SKA project is that being explicit about technical debt helps. We started quite early to log the technical debt formally in our planning, but we did wait to address it, encouraging TDD and refactoring as part of our daily jobs.

What would be your advice for anybody starting to work on a project (big or small and alone or in a team)?

Being explicit about the definition of "done" and the development processes is a key part of maintaining technical debt under control. Adopting test-first practices and allowing for constant refactoring of the code base can be successful techniques in maintaining a high quality of the code base under this aspect. These need to be allowed and encouraged explicitly as part of your development processes and policies. Make well-informed choices, and try to be assessing technical debt proactively: do not wait to discover what is wrong in your system until the last moment. Even the mere knowledge and acceptance of existing technical debt can play a key role in the decision-making process during the development activity. Try not to keep your skeletons in the closet, and be open about the amount of technical debt in your system. Having a well understood process for dealing with it will encourage developers to flag potential issues rather than hiding those.

Note

1. https://www.astropy.org.

Interview: Julien Danjou

Introduction

Julien Danjou is an open-source software veteran who has worked on large projects such as OpenStack and Debian. He currently works at Datadog as staff engineer and is an active participant in many open-source projects including OpenStack. He also runs the pull-request service Mergify.[1] In this interview he tells us about some of his work while at Red Hat.

Summary and Key Insights

In this interview, Julien shows us that the project he worked on tried to adopt most of the best practices. He explains how the project used unit and integration testing (see chapter 6) but illustrates the tradeoffs inherent to technical debt, since not every possible test could be run, given the need for tests to run in limited amounts of time. Later, he also talks about the use of continuous integration systems to merge new code (see chapter 7). In the OpenStack project, most of the debt originated from overengineered features or untested legacy code, often committed by engineers that left the project (an illustration of the impact social factors, chapter 10, might have). Julien emphasizes keeping things simple: do not write code you do not need and avoid over-engineering your product. In other words: keep it simple, stupid.

Detailed Interview

Please introduce yourself, including a brief description of your background and experience in software engineering.

My name is Julien Danjou. I [. . .] work at Red Hat as a principal software engineer. I've been doing open-source development for the past twenty years. For the last seven years, I mostly contributed to OpenStack, the largest Python-based, open-source software project out there. OpenStack is a cloud computing platform providing infrastructure as a service. Most of my contributions to this project were around telemetry. I created and led a few projects in this ecosystem: among them Ceilometer, a metric retrieval system, and Gnocchi, a distributed time series database.

Can you tell us how do you define technical debt and how it applies to your project?

I would define technical debt as code that you know needs to be modified or fixed because it will either break or block you in the future. You have to pay for that code update—with time and some code churn. Your archaic code starts being a debt as soon as you acknowledge it won't be satisfying in the long term, don't act on it, and let it rot in your code base.

The way it applies to the projects I work on is usually coming in different ways. It can be that we do some internal changes and keep backward compatibility on some internal API, and we need to adjust other parts of the code base. Quite often, it can also come from an external dependency: it could change its API, become orphaned or superseded by a new library.

Finally, the worst case is when we take some shortcuts in feature design, knowing that it will block us in the future. This rarely happens, but can be quite problematic when it does.

Tell me about your project: what the project does, how many users, the types of users (individual vs. company), is it open-source or closed-source, any special/challenging requirements, and so on.

Ceilometer is a metric collection system: it retrieves data from other cloud components and stores them into Gnocchi, an open-source distributed time series database. As those projects are open-source, it's difficult to know how many users there are, but we know for sure there's quite a few! They mainly are companies running cloud infrastructure and storing large volumes of metrics.

Ceilometer is really OpenStack-centric, whereas Gnocchi has been designed independently and to be easy to deploy and operate—while being able to work at a very large scale. Its architecture makes it quite unique for a time series database, as all its components are able to scale horizontally.

Tell me about the history of the project: When did it start? Who maintained it a few years back and now? How has the project been architected and managed over time?

Ceilometer started in 2012, and the development team evolved a lot during the early years. As OpenStack gained more and more traction in its first years, we've seen our development team grow very quickly from three maintainers to more than ten—not counting casual contributors. Irregular contributors have been one of our biggest challenges. In an open-source project, developers have different levels of implications, depending on their goals. Some are coming for the long run while some others are just dumping their code into the repository—because they're asked to do so. It's sometimes difficult to see the difference at first sight. However, this latter type of contributor is more dangerous in terms of technical debt: they're unlikely to stick to maintain their code. After the OpenStack hype curve started to decline, a few years ago, our team saw its number of member shrink drastically to three people. This brought a brand new challenge for us, as we were being swamped by a large and outdated code base built by a larger number of developers.

Tell me about the language: What programming languages are used? How do they interact together?

We essentially use Python for all our projects, tools, and libraries.

How is the system tested? Do you use unit testing? Integration testing? Code coverage? What has been used since the inception of the project, and what testing has been added later?

We use several layers of testing, from unit testing to integration testing.

All projects have unit tests, and they are required as part of the merge policy. Reviewers are expected to check that new code is properly covered by unit tests before approving any change.

There are also functional tests, which test the project at a higher level. They are set up by deploying the software with different configuration scenarios. The functional tests are split into different usage situations, and all the scenarios are run before any patch gets merged.

The last piece of testing is integration testing, where all the concerned projects we maintain are deployed in a few different configuration scenarios. Another set of tests is run to check that everything works flawlessly together and that the interaction between the different components isn't broken. Those scenarios are far from being complete or perfect. There are many trade-offs that have been made in order for the test matrix not to be too wide and run in a timely fashion. All those tests scenarios are executed for any pull request that is being sent to the projects—making sure that the patch won't break anything. However, it takes a lot of computing time to run all those checks, so the number of situations that are being tested has to be limited. Resources nor time are infinite.

What would you describe as technical debt on your project? What are the biggest debt points, and how does the team approach and address them? Can you quantify the impact of these debts?

In Ceilometer, most of our technical debt consists of features that are over-engineered, untested, or unused—sometimes all at once. As I mentioned earlier, a large number of the maintainers quit the team over the years and left the remaining maintainers with the burden of supporting the code base. When none of the maintainers have interest nor knowledge in a feature, the code can actually become a burden. A large amount of the development cycle was spent resolving that technical debt. Marginal features were deprecated and then removed; sometimes removing code is actually the easiest way to shrink your debt. On the other hand, widely used functionalities were taken care of and issues were resolved by the maintainers. The fact that a development team has to spend a large amount of its time deprecating features, cleaning code, and upgrading API has a wide impact on the number of new features that it can deliver. There's always compromises to do; in our case, we often picked technical debt resolution over new code, due to the shrinking size of the team. That might pose a problem, as it can also prevent the project from getting interest from new contributors. Yet, those new contributors you don't have could help you maintain the project. Tough calls.

Tell us a story of how technical debt affected your project: What were the impacts and the outcomes?

When the Ceilometer project started, the list of features that users wanted was very long. This made us build a very modular approach for the overall architecture. However, as the project progressed, we eventually realized that some of the pieces were barely used. Moreover, they were a pain to maintain and a bottleneck to the project operational performance. Over several release cycles, the code was marked as deprecated and soon to be removed, to be finally deleted, prompting the architecture to be less complex. This had a tremendous impact on many sides for us. For once, the code started to be simpler to maintain as the code footprint was reduced significantly. Then, the bottleneck being entirely removed, the data processing speed became faster by a large magnitude.

Looking back at your project, what you could have done better in terms of technical debt: prioritization, refactor, and so on?

There are certain parts of our software where we should have been more aggressive, or actually prevented the technical debt from being introduced in the first place. It's not always easy to know in advance what will bite you in the future, but sometimes your experience will give you a good hint. Resolving technical debt can be easy and fun if you know why you do it. It is also an excellent entry point for newcomers that are looking to help your project—especially in open source. I feel like having a better roadmap in certain cases would also have avoided us spending time on useless refactorization. Spending a couple of weeks on refactoring code that is then removed a couple of months later is not what I'd call time well spent. Just skip it.

What would be your advice for anybody starting to work on a project (big or small and alone or in a team)?

If you're alone, be honest with yourself and don't let things rot for too long: you'll be the first (and only) one with a ton of refactoring to do. This can actually lead to the second-system syndrome, because you'll be tempted to restart from scratch, losing most of what you invested so far. If you're part of a team, don't let the same people always sacrifice their time to do the refactoring. Supporting code quality should be a matter shared by everyone. I admit I work with genuine people most of the time; people that take pride in maintaining their part of the project in a tidy state—meaning they clear their debt on a regular basis autonomously. You should target to have collaborators acting this way. If everyone's not playing the game, there are two possibilities. Either your project is too deep in debt, everybody is discouraged, and a sustained campaign to address the issue is needed. Or it is that your team is not engaged in the project for whatever reason (is it boring?) and is just dumping code into your repository without concern.

Obviously, the bigger your code footprint is, the harder it is to maintain a low debt ratio. There's only one [piece of] advice that works in the long term to avoid being weighed down under tons of code churn: don't write more code than you need. You should be utterly lazy and never ever write more code than is strictly required. The more code you produce, the more debt you will have biting you in the future. Leverage libraries whenever possible. Using external products does not mean you won't lose control over the code. If the library is the responsibility of another team, you can probably talk to them. If it's an open-source project, you can send them patches and share maintenance duties.

Note

1. https://mergify.io.

Interview: Nicolas Devillard

Introduction

Nicolas Devillard is an engineer turned product manager with twenty-five years of experience. Through his career, Nicolas has worked with multiple systems and created complex software that runs on heterogeneous platforms. His specialty is with embedded devices, and therefore our interview focuses on this area.

Summary and Key Insights

Nicolas walks us through the challenges developing embedded systems in cryptographic applications. This interview emphasizes that supporting software for a long time requires good design from the start (see chapters 3 and 4); otherwise a change of an upstream dependency can make you rewrite your software almost from scratch. For example, he mentions how early assumptions about direct access to crypto elements are now invalid: "Primitives all assume that they can manipulate crypto key material directly, which is incompatible with most secure elements on the market." This interview also reinforces the need to plan ahead, to manage your debt, and to avoid keeping legacy code with new versions of your product. This is painfully true in the crypto space where your technical debt is a hacker's dream opportunity for exploits.

Detailed Interview

Please introduce yourself, including a brief description of your background and experience in software engineering.

My name is Nicolas Devillard. I studied electronic engineering in Paris with a major in signal processing. Over the past twenty-five years, I have been involved in various roles around software engineering for projects like real-time image processing, astronomical data processing, cryptography, public-key infrastructure, and mobile security. For the past decade I have been acting as a product manager for projects and products related to Internet security.

Can you tell us how you define technical debt and how it applies to your project?

You never build software in a void, right? Even the most esoteric code has to interact with hardware, users, mass storage, networks, and so on. You want to focus on just what your software does and outsource interaction with the outside world to other pieces: an operating system handles interactions with hardware, libraries handle GUI, file parsing, and so on.

Every time you interact with software you have not written yourself, you play the same game: choose an interface, a version number, and program against it. Every single one of these points of contact is also a point of friction that generates technical debt.

The most obvious example of that kind of debt would be graphical user interfaces. When I first started as a young software engineer you had to choose between several candidates: Windows was 3.1, Apple had System 7, SunOS came with SunView or OpenWindows based on X11, other Unix-based systems later came with CDE (Common Desktop Environment) based on the Motif libraries. Pick any one of those and you could be sure you would have to throw everything away and rewrite from scratch a couple of years later.

If you were lucky enough to be in a company that tolerated it, you would design your software into two separate pieces: the GUI, and the inner engine that should survive GUI changes. Unfortunately, most SDK at that time would help you generate skeleton apps for GUI and expect you to write application code inside generated functions, tying them very tightly to that GUI.

I have seen several companies trying to protect against that by asking their engineers to encapsulate GUI layers underneath a proprietary layer. That did not work. A GUI library is not just a set of functions, it also comes with design principles that vary wildly from one platform and language to another, making that proprietary layer impossible to maintain very quickly.

The solution we found at that time (by the mid-1990s) was to switch over to Tcl/Tk for user interfaces. The result was ugly everywhere, but it had the same look and feel and worked everywhere the same. And then we had friction issues with Tcl/Tk and had to make sure we chose one particular version and stuck to it all the way. Technical debt was reduced, but it was still present.

There are other kinds of technical debts you can acquire as a project progresses towards maturity: things that are left in a coarse design state because they do not need to be there from the start. When you come back to those points later, you invariably find out that they turn out to be a lot more complicated than you thought, causing major redesigns that ripple all across the rest of your product.

There is a very thin line between designing software so that you do not close doors to future evolutions and overengineering a piece of code, adding so many useless abstraction layers that you completely lose sight of your initial goals.

Let me illustrate that: you need to save large amounts of data and decide to link against a DB engine. You know that DB engines are just as volatile as fashion these days, and you will probably have to change at some point in the future, so you try to make sure you are not embedding anything that is specific to that engine. But on the other hand, if you do not make use of the strong and unique points that made you choose that engine to start with, you are probably missing something. There is no good solution to that problem, you just need to be aware that every time you use a proprietary feature it will probably make you more efficient for now, but you have just probably created technical debt for the poor soul who will have to migrate to a different engine years down the road.

Tell me about your project: what the project does, how many users, the type of users (individual vs. company), is it open-source or closed-source, any special/challenging requirements, and so on.

The main project I am involved with today is an embedded C library providing TLS connections to tiny devices. The project is open-source, works on pretty much any architecture and OS. You can see it live on GitHub, use it in commercial projects without having to pay a license.

Tell me about the history of the project: When did it start? Who maintained it a few years back and now? How has the project been architected and managed over time?

The software was bought from a small company doing an SSL library under a commercial license. The library was later open-sourced under a permissive license and placed it on GitHub for all to use and contribute. Today the team is mainly located in Cambridge, UK, with help from engineers in France, Poland, and Israel.

The project has grown from a couple of maintainers to a full team, which has not been painless. We recently hit a wall in terms of support for crypto hardware and also wanted to simplify the network layer, so we went into a major reengineering of the whole stack over the past year or so. The new version should break all APIs but also provide the foundations for more generic crypto support in all embedded devices.

Tell me about the language: What programming languages are used? How do they interact together?

This is all C, with the odd Makefile and Python scripts to help build the library. Interactions with the underlying OS are fairly limited and abstracted to help porting.

How is the system tested? Do you use unit testing? Integration testing? Code coverage? What has been used since the inception of the project, and what testing has been added later?

All crypto primitives are tested using standard test vectors and then some. We also test compatibility against other TLS stacks. Most important for us is that the library can be built and used on tiny Cortex-M microcontrollers, which make the bulk of those tiny smart devices surrounding us. We have access to a farm of microcontroller boards and can test the library runs the same way everywhere.

What would you describe as technical debt on your project? What are the biggest debt points, and how does the team approach and address them? Can you quantify the impact of these debts?

The biggest debt we have is support for the legacy APIs we have today. Our users have been [relying] on those APIs for years and changing anything is likely to break a lot of existing software.

The original software was designed as a small embeddable library taking care of SSL/TLS connections to the outside world. Since those heavily rely on crypto primitives, we have a whole collection of algorithms like AES, DES, RSA, and elliptic curves that have been developed purely for this library. But then we found out that most users of the library are more interested in using the crypto primitives than the TLS features themselves. Most of the feedback we are receiving is about making those primitives stronger, smaller, faster, which most often requires hardware support. Unfortunately the legacy library is not designed for that. Primitives all assume that they can manipulate crypto key material directly, which is incompatible with most secure elements on the market.

We had to rethink the whole model behind crypto: instead of dealing with crypto keys directly, developers are expected to refer to key by handles and use the library as a black box offering crypto services. You can use the keys but can never see them directly. Saves a lot of ways you could shoot yourself in the foot, and also opens the gates to supporting secure elements like smart cards. It also allows us to work with partitioned environments—for example, TrustZone on Cortex-A and equivalent beasts on Cortex-M. One partition would hold the standard firmware, and another partition would hold the crypto keys and only offer access to them through operations—for example please sign this bag of bytes with the device private key and give me the result, or please decrypt those bytes and send me the clear text but don't tell me the key.

Tell us a story of how technical debt affected your project: What were the impacts and the outcomes?

Legacy code would assume that crypto keys are moved around as part of a C struct. The reengineered version forces the caller to identify a key handle and request an operation to be performed. This adds a layer of abstraction for key protection, which is nice for security but has a price in terms of code size and speed, since you are always a couple of function calls away from the real crypto operations. In some cases this increased the library size beyond what can fit a tiny microcontroller.

Our hope is that we make it so easy to use crypto hardware that all equivalent software crypto primitives can be compiled out and replaced by calls into hardware drivers, reducing the overall footprint.

The result is not black or white: we now have a more secure library, which is really the point of embedding a protocol like TLS in the first place, but we have also increased our requirements in terms of code size and use of resources. We also broke a lot of software along the way, though this will be mitigated by supporting legacy versions for several years to give customers enough time to upgrade.

Doing nothing was not an option. There are already several successful academic and practical attacks against the legacy code we have. Without the latest reengineering we would have gone straight into a wall, and the library would have ended as a piece of software with known security holes that are too expensive to patch.

Reimplementing an existing piece of software is a lot more work than it seems. As a software engineer faced with a wall of legacy tech, you may be thinking, "Gee, these are thirty-six classes dedicated to doing something that is natively provided in Go. There are countless bugs in there and I am not even sure I understand all the details. As soon as we move to Go this is all going away."

This is true, but you forgot that the legacy tech is in use by 200 customers around the globe. Come over with your shiny new Go version and they will ask you, "What's in it for me?" If your new version offers the same features, your customers have no reason to upgrade, and you will end up having two versions to maintain: the legacy one you wanted to get rid of and the new ones with its new bugs.

In a previous company we had planned to replace four legacy products with just one, but ended up with 4 + 1 products: the legacy versions and the new product. Debt is sticky, it can stay for extended periods of time before it can be replaced with shiny new tools. There is a reason why the banking world is still relying on heaps of COBOL software.

Looking back at your project, what you could have done better in terms of technical debt: prioritization, refactor, and so on?

The initial effort for refactoring the library from direct key access to key store mode was estimated to three to six months. We are now eighteen months later, and the effort is still not completed. We are effectively maintaining legacy versions in parallel with the reengineered version, which takes an enormous toll on the team and slows everything down. We tried adding engineers but there is no magic here: the new engineers need training by the existing team, which stops all efforts on the project for a while.

Going longer term, we know we will have to suffer from maintaining legacy and new versions simultaneously for a few years. If there is one thing I could change by going back, it would be motivating our managers to hire more people upfront for this effort. I clearly underestimated the amount of work needed to realize that transformation.

What would be your advice for anybody starting to work on a project (big or small and alone or in a team)?

The main lesson I learned is that you cannot get it right on the first attempt. In The Mythical Man-Month, Brooks said, "plan to throw away the first version because that is what is going to happen anyway." The best way I found to get past that stage is to start with prototypes written in a permissive language that allows you to focus on your project. I typically use Python or Go, then learn from the prototype and start over in another language. You can easily throw away a few pages of Python, but you will have a hard time deleting precious C++ classes and C functions that have taken you painful evenings to write in full concentration. Make sure your model works before you spend all of your energy and creativity into implementing it.

I have seen countless examples of engineers heading straight into C++, producing mountains of code and classes, realizing a bit late that the whole model could not possibly work. Their reaction was invariably to bend the model and the user's will to make it fit what they had already programmed. That is a sure way to produce software nobody will ever want to use, and then marketing blames the market for not being ready and you got yourself a good excuse for a disaster—but in reality you have been trying to ship a version 1 which was supposed to be a learning exercise ending in the bin.

One criterion I use to judge of the health of a project is by listening to engineer discussions at coffee time. Are they endlessly debating about the tools they are using, or are they talking about the product they are building? If lunchtime and coffee sessions are dedicated to heated fights around C++ templating or compiler options rather than satisfying customer needs, you can be sure the team is already too deep down the rabbit hole to remember what the whole point of the project was.

Interview: Vadim Mikhnevych

Introduction

Vadim Mikhnevych is a software engineer and tech lead at SoftServe. He has worked primarily in Java since 2005 in telecom and healthcare projects. With more than sixteen years of experience in software development, he has seen his fair share of technical debt.

Summary and Key Insights

This interview emphasizes that a system's software architecture (chapter 4) is a major pillar of the project, and that performance issues can force developers to refactor an application to meet performance goals. Issues are amplified by a lack of testing (in this case, a lack of performance testing), which reinforced the need to manage testing debt, as discussed in chapter 6. Vadim also illustrates how team management and social debt, including global software development challenges, impact product quality, as we discuss in chapter 10. Finally, close attention to quality attribute requirements early on (chapter 3) is important to understand tradeoffs the system is making.

Detailed Interview

Please introduce yourself, including a brief description of your background and experience in software engineering.

Vadim Mikhnevych, tech lead. I've been working with Java since 2005, and the last six years at SoftServe. Mostly telecom and healthcare projects.

Can you tell us how do you define technical debt and how it applies to your project?

That some job [is] left undone because it does not result directly into delivering some features, and therefore is postponed because of other tasks having higher priority.

Tell me about your project: what the project does, how many users, the types of users (individual vs. company), is it open-source or closed-source, any special/challenging requirements, and so on.

Ambulatory information system, in other words—another EHR application. Number of users is not defined, users are company-based (it should be a multitenant application with each customer company having its own set of users). Closed-source. Most challenging requirements are the changing ones.

Tell me about the history of the project: When did it start? Who maintained it a few years back and now? How has the project been architected and managed over time?

The project started about three years ago, I switched to it two years ago after the acting BE tech lead left the company. Currently I'm the longest working person from SoftServe on this project—the original team left one by one during these years, and new people came. Since that time, the project expanded, and currently it is developed not only by SoftServe but also by customer teams from the USA, Romania, and other EU countries. Most architecturing and management is done on the customer side, however, there is a regular architectural meeting with techlead-level representatives of all the teams, where decisions are proposed and discussed.

Tell me about the language: What programming languages are used? How do they interact together?

Java backend interacts with JS/Angular frontend via REST services. There are some additional server-side components written on NodeJS, like audit, which are communicated with over HTTP.

How is the system tested? Do you use unit testing? Integration testing? Code coverage? What has been used since the inception of the project, and what testing has been added later?

All the mentioned kinds of testing was present. Though code coverage was initially very low, lately it improved.

What would you describe as technical debt on your project? What are the biggest debt points, and how does the team approach and address them? Can you quantify the impact of these debts?

Most noticeable debt lies in areas on maintainability and performance. There are some performance issues which need refactoring and affect everyday testing and development, but which are not considered of high priority. We also lack performance testing. Regarding maintainability, there is some amount of duplicated or boilerplate code. This has low impact and is addressed by the team occasionally, using "boy scout rule" [the boy scout rule says to leave your campsite cleaner than you found it—NE]

Tell us a story of how technical debt affected your project: What were the impacts and the outcomes?

Recently, lack of attention to performance caused snail-speed demo and somewhat irritated the customer higher-level representatives. This caused the priority of these issues to be increased, which is good.

Looking back at your project, what you could have done better in terms of technical debt: prioritization, refactor, and so on?

I would dedicate about 25% of development time to nonfunctional requirements, including also refactoring and technical debt management.

What would be your advice for anybody starting to work on a project (big or small and alone or in a team)?

Don't leave nonfunctional requirements for later.

Interview: Andriy Shapochka

Introduction

Andriy Shapochka is a veteran software engineer with twenty-three years of experience. He is currently a distinguished architect at SoftServe. He has worked with many languages, operating systems, and software domains. Currently he provides consulting expertise to internal and client projects with SoftServe, working with Java, Python, and Javascript.

Summary and Key Insights

This interview covers technical debt in popular business development languages and some of the tools Andriy has seen applied, including the use of SonarQube for data analytics. During our discussion, he is clear that technical debt incurs real financial costs, but stakeholders still misunderstand its consequences and are still reluctant to invest in its remediation (see chapter 11). This interview also confirms that managing technical debt is a constant effort and it requires 15% to 20% of the project's total engineering effect to keep the interest down at a reasonable level. Finally, Andriy strongly believes that some initial effort on building the architecture underpinnings is essential to avoiding longer-term technical debt, as we discuss in chapter 4.

Detailed Interview

Please introduce yourself, including a brief description of your background and experience in software engineering.

Andriy Shapochka, distinguished architect at SoftServe, has twenty-three years of experience in software engineering and architecture overall, including design for the new software systems and assessment of the existing ones.

Can you tell us how do you define technical debt and how it applies to your project?

In simple words, technical debt grows from the decisions made or unmade, typically, to speed up feature implementation and delivery in the short run that negatively affect team development velocity and quality in the long run (add friction slowing project evolution).

Tell me about your project: what the project does, how many users, the type of users (individual vs. company), is it open-source or closed-source, any special/challenging requirements, and so on.

I typically do not work with long-term projects as an architect; rather I am involved in consulting, including assessments for projects run by other teams (SoftServe or third party). All of them are closed-source. Most of the projects I deal with have a large code base ranging from hundreds of thousands to millions LOC [lines of code]. They are frequently complex, convoluted, and over engineered with the clear signs of rotten code.

Tell me about the history of the project: When did it start? Who maintained it a few years back and now? How has the project been architected and managed over time?

Two typical cases are either the project is maintained by the same organization from the very beginning with the team growing and rotating its members, or the initial project implementation gets owned as part of another company's acquisition, and the acquiring company continues to maintain and evolve the codebase. In most cases knowledge of the original architecture is either lost or out of sync with the actual state of implementation. Typical project duration is at least a few years.

Tell me about the language: What programming languages are used? How do they interact together?

I mostly work with Java, Python, JavaScript stacks. Front-end is typically implemented with JavaScript and integrates with back-ends coded in one of those stacks via REST API.

How is the system tested? Do you use unit testing? Integration testing? Code coverage? What has been used since the inception of the project, and what testing has been added later?

As a rule unit, integration, and load testing are implemented to some degree, but since project sponsors are reluctant to commit resources to tests which obviously do not directly add up to the features of the implemented system and are not understood by the decision makers the tests are frequently insufficient, out of date, or do not bring much value, being too simplistic to properly test business logic, complex scenarios, extreme loads.

What would you describe as technical debt on your project? What are the biggest debt points, and how does the team approach and address them? Can you quantify the impact of these debts?

Most teams look at the SonarQube analytics as the source to discover and quantify technical debt. Later on they discover that addressing issues found by SonarQube does not necessarily contribute to essential debt elimination. The real debt causing most friction in the project is not captured in the tools or documentation in many cases. I think the most useful quantification of the technical debt could be represented as a relative cost of change required to evolve a specific part of the system to make it support new requirements, changing requirements, postponed requirements (can be functional or non-functional). The word "relative" here means comparison of what it takes to change the actual code with the would-be cost of change in assumption the code is free from the identified technical debt item(s).

Tell us a story of how technical debt affected your project: What were the impacts and the outcomes?

The most frequent impact caused by the technical debt is steep growth in the cost of project evolution combined with the low team development velocity. Technical debt also affects the number of defects existing in the codebase and hinders their troubleshooting and elimination. The side effect is also that people (esp. senior level engineers) are reluctant to join projects ridden with technical debt.

Looking back at your project, what you could have done better in terms of technical debt: prioritization, refactor, and so on?

One feasible approach here is to regularly commit a certain percentage of development time (15% or 20%) to work on technical debt—its analysis, prioritization, and incremental refactoring. It is also useful to schedule regressions and stabilization before public releases of the product versions.

We should always try to pay the debt off before the interest on it has gone to the sky. . . .

What would be your advice for anybody starting to work on a project (big or small and alone or in a team)?

Unless it is a throw-away prototype, try to start not with features but with the cross-functional foundation: support for proper authentication and authorization, logging, efficient integration with databases, other data sources and message queues, proper API definitions, horizontal scalability support, solid choices for frameworks and libraries. This will put the project on the straight rails from the beginning and will help avoid many types of technical debt (especially architectural) from the start.

Index

Note: Page numbers in *italics* denote figures and other illustrations and those in **bold** type denote tables.